PLANT RESISTANCE TO VIRUSES

The Ciba Foundation is an international scientific and educational charity. It was established in 1947 by the Swiss chemical and pharmaceutical company of CIBA Limited—now CIBA-GEIGY Limited. The Foundation operates independently in London under English trust law.

The Ciba Foundation exists to promote international cooperation in biological, medical and chemical research. It organizes about eight international multidisciplinary symposia each year on topics that seem ready for discussion by a small group of research workers. The papers and discussions are published in the Ciba Foundation symposium series. The Foundation also holds many shorter meetings (not published), organized by the Foundation itself or by outside scientific organizations. The staff always welcome suggestions for future meetings.

The Foundation's house at 41 Portland Place, London, W1N 4BN, provides facilities for meetings of all kinds. Its Media Resource Service supplies information to journalists on all scientific and technological topics. The library, open seven days a week to any graduate in science or medicine, also provides information on scientific meetings throughout the world and answers general enquiries on biomedical and chemical subjects. Scientists from any part of the world may stay in the house during working visits to London.

PLANT RESISTANCE TO VIRUSES

A Wiley – Interscience Publication

1987

JOHN WILEY & SONS

Chichester · New York · Brisbane · Toronto · Singapore

© Ciba Foundation 1987

Published in 1987 by John Wiley & Sons Ltd, Chichester, UK.

Library of Congress Cataloging in Publication Data
Plant resistance to viruses.

(Ciba Foundation symposium; 133)
"A Wiley–Interscience publication."
Editors, David Evered (organizer) and Sara Harnett.
Symposium held at the Ciba Foundation, London,
31 March–2 April 1987.
Includes indexes.
1. Virus diseases of plants—Congresses. 2. Plants—
Disease and pest resistance—Congresses. 3. Plant
viruses—Congresses. I. Evered, David. II. Harnett,
Sara. III. Series.
SB736.P57 1987   582'.0234   87–25447
ISBN 0 471 91263 8

British Library Cataloguing in Publication Data

Plant resistance to viruses.—(CIBA
   Foundation symposium; 133).
   1. Virus diseases of plants 2. Plants—
   Disease and pest resistance
   I. Series
   581.2'34   SB736

   ISBN 0 471 91263 8

Typeset by Inforum Ltd, Portsmouth
Printed and bound in Great Britain by the Bath Press Ltd., Bath, Avon.

Contents

Participants

J.F. Antoniw   Crop Protection Division, Plant Pathology Department, Rothamsted Experimental Station, Harpenden, Herts AL5 2JQ, UK

D.C. Baulcombe   Plant Breeding Institute, Maris Lane, Trumpington, Cambridge CB2 2LQ, UK

R.N. Beachy   Department of Biology, Washington University, Campus Box 1137, St Louis, Missouri 63130, USA

J.F. Bol   Department of Biochemistry, State University of Leiden, Wassenaarseweg 64, 2333 AL Leiden, The Netherlands

G. Bruening   Department of Plant Pathology, University of California, Davis, California 95616, USA

J.W. Davies   John Innes Institute, Colney Lane, Norwich NR4 7UH, UK

J.A. Dodds   Department of Plant Pathology, University of California, Riverside, California 92521, USA

J.E. Duffus   Sugarbeet Production Research Unit, US Agricultural Research Station, 1636 East Alisal Street, Salinas, California 93905, USA

R.S.S. Fraser   Institute of Horticultural Research, Worthing Road, Littlehampton, BN17 6LP, UK

B. Fritig   Laboratoire de Virologie, Institut de Biologie Moléculaire des Plantes du CNRS, 12 Rue du Général Zimmer, 67000 Strasbourg, France

S. Gianinazzi   INRA, Station de Génétique et d'Amélioration des Plantes, BV 1540, F-21034 Dijon cédex, France

R.W. Goldbach   Department of Virology, Agricultural University, Binnenhaven 11, 6709 PD Wageningen, The Netherlands

B.D. Harrison (*Chairman*)   Virology Division, Scottish Crops Research Institute, Invergowrie, Dundee DD2 5DA, UK

T. Hohn   Friedrich Miescher-Institut, PO Box 2543, CH-4002 Basel,
Switzerland

G. Loebenstein   Department of Virology, Institute of Plant Protection,
Agricultural Research Organization, The Volcani Centre, POB 6,
IL-50250 Bet Dagan, Israel

S.A. Lommel (*Ciba Foundation Bursar*)   Department of Plant Pathology,
Kansas State University, Throckmorton Hall, Manhattan, Kansas 66506,
USA

K. Maramorosch   Department of Entomology, Rutgers – The State
University, PO Box 231, New Brunswick, New Jersey 08903, USA

R.E.F. Matthews   Department of Cell Biology, University of Auckland,
Private Bag, Auckland, New Zealand

M. Nishiguchi   National Institute of Agrobiological Resources, Tsukuba
Science City, Yatabe, Ibaraki 305, Japan

E. Rybicki (*Ciba Foundation Bursar*)   Microbiology Department, Molecular
Biology Building, University of Cape Town, PB Rondebosch 7700,
South Africa

H.L. Sänger   Max-Planck-Institut für Biochemie, D-8033 Martinsried bei
München, Federal Republic of Germany

I. Sela   Virus Laboratory, The Levi Eshkol School of Agriculture, The
Hebrew University of Jerusalem, POB 12, IL-76100 Rehovot, Israel

J.L. Sherwood   Oklahoma State University, Department of Plant Pathology,
Division of Agriculture, Life Sciences East 104, Stillwater,
Oklahoma 74078-0285, USA

J. van Emmelo   Plant Genetic Systems n.v., Laboratories, J Plateaustraat 22,
B-9000 Ghent, Belgium

A. van Kammen   Department of Molecular Biology, Agricultural
University, De Dreijen 11, 6703 BC Wageningen, The Netherlands

L. van Vloten-Doting   Research Institute Ital, PO Box 48,
6700 AA Wageningen, The Netherlands

R.F. White   Crop Protection Division, Plant Pathology Department, Rothamsted Experimental Station, Harpenden, Herts AL5 2JQ, UK

M. Zaitlin   Department of Plant Pathology, Cornell University, 334 Plant Science Building, Ithaca, New York 14853, USA

D. Zimmern   MRC Laboratory of Molecular Biology, Hills Road, Cambridge CB2 2QH, UK

Introduction

B.D. Harrison

Virology Division, Scottish Crop Research Institute, Invergowrie, Dundee, UK

1987 Plant resistance to viruses. Wiley, Chichester (Ciba Foundation Symposium 133) p 1–5

A symposium of this kind needs some introductory comments to set the scene and to point to topics that seem especially to merit consideration. Let me first emphasize that our discussions will take place against a background of the knowledge that virus diseases still cause large yield losses in many crop species. Therefore, progress in solving the scientific problems posed by the need to increase virus resistance in crop plants has important implications for agriculture.

Crop losses attributable to virus diseases are greatest in, but by no means confined to, areas with warmer climates that favour the reproduction and activity of virus vectors. They are especially serious in developing countries in the sub-tropics and tropics. In tropical Africa, for example, maize streak and groundnut rosette viruses can cause devastating disease epidemics, and cassava mosaic is responsible for crop losses which are conservatively valued at more than two hundred million pounds per annum. Over the years, three main categories of control measure have been adopted for preventing virus-induced crop losses. The first type aims to remove virus sources, for example by producing virus-free planting stocks of vegetatively propagated plants, or by removing volunteer plants or plant propagules left from previous crops. The second type is concerned with preventing virus spread, usually by killing or interfering with the activity of virus vectors. The third type, which is the most economical for farmers, is to grow virus-resistant varieties of crops.

In this symposium we are concerned only with virus resistance. This choice seems both appropriate and timely for two main reasons. The first is that the environmental consequences of applying large amounts of vector-killing pesticides to crops are becoming increasingly evident, and public pressure has grown, and can be expected to continue to grow, for their more restricted and more discriminating use. A further factor is that, in some areas where pesticides have been widely used, pesticide-resistant vector organisms have become common (Table 1). A consequence of these developments is that the scope for using crop protection chemicals to prevent virus spread may decrease. The second reason is that recent advances in molecular biology and

TABLE 1   Increases in insecticide resistance of hopper vectors of rice viruses in Taiwan

Insecticide	Increase in insecticide resistance[a]	
	Nephotettix cincticeps (green rice leafhopper)[b]	*Nilaparvata lugens* (brown planthopper)[c]
Malathion (organophosphate)	× 27–452	× 288–526
Carbaryl (carbamate)	× 12–79	—
Propoxur (carbamate)	—	× 19–46
Permethrin (pyrethroid)	× 1–6	× 71–121

[a] Range of factors of increase in insecticide resistance of field populations relative to a susceptible control population.
[b] Data from Kao et al (1982).
[c] Data from Chung et al (1982).

biotechnology offer the prospect, not only of speeding up progress in the more conventional approaches to producing virus-resistant plants, but also of exploiting completely new approaches.

Interest in the resistance of varieties of plants to virus diseases probably goes back for at least a couple of centuries, to a time that pre-dated the recognition of viruses as a separate class of pathogens. Although the correct interpretation of old writings can be debatable, an apparently relevant observation is to be found in a book by Marshall (1790) on the rural economy of the Midland counties of England. Marshall recognized varietal differences in potato in the occurrence of 'curledtop', the disease that we now attribute to infection with potato leafroll virus. He wrote: 'The old varieties, formerly in cultivation, dwindling in produce, and being, at length, in a manner des-troyed, by the disease of curledtop, two new varieties were introduced. . . . The consequence has been, the disease vanished with the old sorts, and is now (1786) and in *this* neighbourhood, where no other sort is in ordinary cultiva-tion, in a manner forgot'. Although it is uncertain whether the events re-corded by Marshall represented the replacement of infected stocks with virus-free clones of other cultivars, or with virus-resistant cultivars, it seems unlikely that all the old infected stocks would have been discarded in the same year, and the virus reservoirs therefore removed simultaneously, and more probable that the new varieties were much more resistant to, or tolerant of, infection than the old ones. This kind of selection for superior performance in the field has proceeded, unconsciously as well as consciously, for centuries. As a result many of the varieties of crop plants in cultivation in areas where these crops have been exposed to a prevalent virus for a long period are cultivars which are tolerant of infection. This is found, for example, in native potato genotypes grown in the Andean region of Peru (Jones 1981). Similarly, the earliest attempts to breed improved crop plants relied on selection, now more often intentional, to eliminate the most readily infectible and sensitive types.

This first phase of plant improvement was succeeded by one in which attempts were made to breed virus-resistant forms by selecting and crossing appropriate parents, and then making selections from among their progeny, backed, where possible, by knowledge about the genetic control of resistance. Good progress was sometimes made without detailed knowledge of the genetic control, as instanced by the programme of breeding sugar beet for resistance to curly top in the United States (Carsner 1933).

Where the range of genetic variation found in a crop species does not include the required degree of virus resistance, this can sometimes be identified in a related species. Efforts to introduce the resistance genes into such crop plants are a feature of the third phase of breeding for virus resistance. For example, the R_y gene for extreme resistance to potato virus Y (Ross 1960) has been transferred by breeders from the primitive species *Solanum stoloniferum* to *S. tuberosum*. Once introduced into suitable parental material, such dominant genes are relatively easy to include in breeding programmes. Recessive genes can also be of value, as exemplified by the resistance of groundnuts to groundnut rosette disease, a property thought to be controlled by two recessive genes (K.R. Bock & S.M. Nigam, unpublished results). The growing knowledge of the genetics of resistance is described in detail by Fraser in this symposium: he considers, on the one side, the genetics of resistance in the plant and, on the other, the genetics of virulence in the virus.

Recent research has built on these foundations in two main ways. First, modern techniques (including electron microscopy, protoplast methodology, and biochemical and serological analysis) have enabled resistance mechanisms within a species to be examined at the cellular and cell-free levels as well as in tissues and intact plants. As a result of these and other analyses, it has become clear that resistance mechanisms can be assigned to two principal categories: innate resistance and induced resistance. *Innate resistance* is heritable and constitutive, and can take a range of forms, some of which are becoming much better understood (see Bruening et al and Nishiguchi & Motoyoshi, this volume). *Induced resistance* is not constitutive and is expressed only after it is activated by some previous infection or treatment. Cross-protection between virus strains is one such example (Sherwood, this volume). Recent information on virus non-specific induced local, and induced systemic, resistances is dealt with in the papers on pathogenesis-related (PR) proteins and antiviral factor (AVF). As an aside, I feel that the designation 'pathogenesis-related' for the PR proteins is unfortunate because these proteins can be produced in substantial amounts without the intervention of a pathogen (Fraser 1981) and they seem more related to stress, or perhaps to ageing, than merely to pathogenesis.

The second important development stems from the application of molecular biological and genetic engineering techniques, and is bringing totally new approaches to virus resistance within our grasp. On the one hand, attempts are under way to isolate virus resistance genes that occur naturally in plants and to define them in molecular terms as a preliminary to transferring them to other plant cultivars or species. On the other hand, we now have examples of at least two successful approaches to enhancing virus resistance by transforming the genome of plants with nucleotide sequences copied from the genetic material of the viruses themselves. In parallel with these genetic engineering approaches to resistance we have the possibility of using viruses to introduce non-viral genetic material into plant cells. These developments, which will be discussed in the last part of this symposium, are the first few small fruits from the application of what is a radically new addition to the range of techniques available for improving plants. As with more conventional kinds of resistance, these novel kinds must be tested carefully for their durability in field conditions, and possible side-effects and environmental hazards must also be assessed.

There is now, therefore, a more impressive array of approaches to improving virus resistance than has been available before, and we may be entering an age in which virus diseases will be controlled as effectively as bacterial diseases are today. However, viral genomes have probably survived and evolved over long periods and it would be surprising if plant viruses lacked the genetic flexibility to generate new forms capable of surviving in a world of plant breeders, virologists and genetic engineers. For the present, I hope that by describing, discussing and drawing together the many recent findings that are relevant to virus resistance, clearer views will emerge both of the scientific problems that are most urgently in need of solution, and also of the experimental approaches that seem suitable for solving them. A whole range of scientific problems come to mind, some more tractable than others. They include the following:

To what extent are kinds of resistance mechanism related to kinds of genetic control in the plant?

How can knowledge about PR proteins and antiviral factors be exploited?

What is the mechanism of cross-protection between virus strains, and does it differ in different virus groups?

How can virus resistance genes in plants best be defined and characterized at the molecular level?

What are the prospects for new or improved genetic engineering approaches to enhancing virus resistance?

How durable in field conditions are different kinds of resistance likely to be?

How do viruses overcome the effects of resistance genes?

What are the main barriers to progress, and how can they be removed?

This personal list is by no means exhaustive but I hope it will serve to provoke thought and discussion when considering the points made by the main contributors to the symposium.

References

Bruening G, Ponz F, Glascock C, Russell ML, Rowhani A, Chay C 1987 Resistance of cowpeas to cowpea mosaic virus and tobacco ringspot virus. In: Plant resistance to viruses. Wiley, Chichester (Ciba Found Symp 133) p 23–37

Carsner E 1933 Curly-top resistance in sugar beets and tests of the resistant variety U.S. No. 1. USDA Tech Bull 360

Chung T-C, Sun C-N, Hung C-Y 1982 Resistance of *Nilaparvata lugens* to six insecticides in Taiwan. J Econ Entomol 75:199–200

Fraser RSS 1981 Evidence for the occurrence of the 'pathogenesis-related' proteins in leaves of healthy tobacco plants during flowering. Physiol Plant Pathol 19:69–76

Fraser RSS 1987 Genetics of plant resistance to viruses. In: Plant resistance to viruses. Wiley, Chichester (Ciba Found Symp 133) p 6–22

Jones RAC 1981 The ecology of viruses infecting wild and cultivated potatoes in the Andean region of South America. In: Thresh JM (ed) Pests, pathogens and vegetation. Pitman, London, p 89–107

Kao H-L, Liu M-Y, Sun C-N 1982 *Nephotettix cincticeps* (Homoptera:Cicadellidae) resistance to several insecticides in Taiwan. J Econ Entomol 75:495–496

Marshall W 1790 Rural economy of the Midland Counties, vol 1. G Nicol, London

Nishiguchi M, Motoyoshi F 1987 Resistance mechanisms of tobacco mosaic virus strains in tobacco and tomato. In: Plant resistance to viruses. Wiley, Chichester (Ciba Found Symp 133) p 38–56

Ross H 1960 Der Praxis der Züchtung auf Infektionsresistenz und extremer Resistenz (Immunität) gegen das Y-Virus der Kartoffel. Eur Potato J 3:296–306

Sherwood JL 1987 Mechanisms of cross-protection between plant virus strains. In: Plant resistance to viruses. Wiley, Chichester (Ciba Found Symp 133) p 136–150

Genetics of plant resistance to viruses

R.S.S. Fraser*

Institute of Horticultural Research, Wellesbourne, Warwick CV35 9EF, UK

Abstract. This paper concerns the genetics of resistance used by the plant breeder to produce cultivars resistant to viruses. Non-host immunity, and resistance induced in normally susceptible individuals, are discussed only where they may share mechanisms with cultivar resistance. Conclusions about the genetics of resistance and of virulence (the ability of a virus isolate to overcome a specific resistance gene) are drawn from a survey of 63 combinations of hosts and viruses, and from comparisons with the predictions made from various theoretical models of host–virus interactions. Most resistance mechanisms that result in virus localization appear to involve an inducible, positive inhibitor of virus replication or spread, which tends to be temperature sensitive. Resistance mechanisms which permit some systemic spread of virus tend to be incompletely dominant (gene-dosage dependent) and are determined by quantitative interactions between host- and virus-specified functions. Completely recessive resistance is rare, and may involve a negative mechanism where the resistant plant lacks a susceptibility function. Most of the resistance genes considered have been overcome by virulent isolates of virus; extreme durability is rare. It appears easier for viruses to mutate to overcome dominant localizing resistance than recessive immunity mechanisms.

1987 Plant resistance to viruses. Wiley, Chichester (Ciba Foundation Symposium 133) p 6–22

Crop losses caused by plant virus diseases can be controlled in various ways. Infection can be prevented by good hygiene, use of virus-free seed and control of vectors. Viruses can be eliminated from universally infected cultivars or clonally propagated lines by tissue culture techniques. However, in the absence of any chemical treatment analogous to the use of fungicides, the only strategy of control that can be applied directly to field crops is breeding for host resistance.

Full exploitation of resistance in crop protection depends on an understanding of the genetical and biochemical mechanisms involved, and of the nature of the plant–pathogen interaction. In this paper I shall summarize knowledge of the genetics of resistance, and of resistance-breaking behaviour in the virus. I shall indicate where genetic information suggests possible mechanisms. Finally, I shall stress some of the limitations of disease control

---

* *Present address*: Institute of Horticultural Research, Worthing Road, Littlehampton BN17 6LP, UK.

by classical breeding for resistance, and speculate on how the approach could be expanded.

Types of resistance

There has been some confusion in the literature over nomenclature, because of the diversity of plant resistance mechanisms and of the corresponding phenomena in the virus. It is therefore useful to begin with some definitions. Resistance mechanisms can be separated into three broad groups, which operate at different levels of complexity of the host population.

In *non-host resistance* all individuals of a species are completely unaffected by a particular virus; on inoculation the virus produces no symptoms or detectable multiplication. Two genetic models can be suggested. In the positive model, the 'non-host' contains a gene or genes completely effective against all tested isolates of the virus. Holmes (1955) suggested as many as 20 to 40 genes with additive effects, but there is no direct evidence for such polygenic resistance systems in plants, and in any case they would be very difficult to handle in breeding programmes. Bald & Tinsley (1967) suggested a negative mechanism; a species is a non-host because it lacks certain 'susceptibility factors' required by the virus for full pathogenesis. Again there is little direct evidence to support this, but a possible site of action would be cooperation of host- and virus-coded subunits to form a functional replicase. 'Negative' mechanisms of non-host resistance could only be exploited in plant breeding by the modification or deletion of some existing host function. Finally, recent evidence shows that protoplasts, isolated from some plants considered to be non-hosts, can support virus multiplication (e.g. Huber et al 1981). This suggests that at least some cases of non-host immunity may be mediated by physical barriers to infection at the cell wall or epidermis.

Cultivar resistance occurs within a host species. Resistant individuals contain a gene or genes conferring resistance to a virus which affects susceptible members of that species. This is the type of resistance most used by the plant breeder, and is the main subject of this paper. The corresponding effect in the virus is *virulence* — the ability to overcome a specific resistance gene and thus cause disease in a resistant plant.

Induced resistance operates at the level of the individual, when a form of resistance is conferred on a susceptible plant by a prior inoculation, or chemical or environmental treatment. It includes effects such as acquired systemic resistance and the pathogenesis-related proteins, cross-protection, and virus-free green islands in mosaic tissue. The mechanisms are diverse; some may depend on host genes directly involved in cultivar resistance mechanisms, while others are probably indirect effects of other aspects of host metabolism (reviewed by Fraser 1987). Unlike cultivar resistance, induced resistance is not normally heritable, and must be conferred afresh on each generation.

TABLE 1 Genetics of resistance to viruses in crop species and some features of resistance gene action and virulence (derived from data in Fraser 1986, 1987)

Genetic basis:

	Number of host–virus combinations
Single dominant gene	29
Incompletely dominant (gene-dosage dependent)	10
Apparently recessive	11
Sub-total: monogenic	50
Possibly oligogenic	5
Monogenic, with possible modifier genes or effects of host genetic background	8
Sub-total: oligogenic (?)	13
Total number of host–virus combinations in sample	63

Localization[a]:

	Immune	Yes	Partial	No	Not known	Total
Dominant alleles	0	19	0	2	8	29
Incompletely dominant	0	0	4	8	0	12
Apparently recessive	5	1(?)	1	2	4	13

Temperature response[b]:

	ts	tr	Not known	Total
Dominant alleles	7	2	20	29
Incompletely dominant	1	1(?)	10	12
Apparently recessive	2	2	9	13

Virulent isolates reported:

	Yes	No	Not known	Total
Dominant alleles	16	1	12	29
Incompletely dominant	8	3	1	12
Apparently recessive	4	1	8	13

[a] Immune, no virus detectable; Yes, normally involving lesion formation; No, resistance permitting some systemic spread; Not known, not tested, or not reported in the literature.
[b] ts, temperature sensitive; tr, temperature resistant.

The genetics of cultivar resistance

Heritable resistance is known in numerous crop species. The numbers of genes involved have been determined by standard genetic methods, by fitting observed segregation ratios to predictions for various models. Resistance alleles have been classified as dominant, incompletely dominant or recessive, depending on the resistance phenotype of plants homozygous or heterozygous for the resistance allele. Table 1 summarizes the genetic control of resistance in a randomly chosen sample of 63 combinations of hosts and their viruses. Fuller details and literature citations are given elsewhere (Fraser 1986).

In these cultivated species, resistance is mostly inherited at a single locus. The evidence for oligogenic control, or modifier genes, is weaker. Some early examples of genetically complex resistance were later shown to be monogenic (reviewed by Fraser 1986). The early experiments were conducted under variable environmental conditions, and the proposed genetic complexity was an attempt to explain, in purely genetic terms, segregation ratios resulting from genotype–environment interactions. There are, however, a few cases of well-substantiated host genes which indirectly modify the phenotypic expression of resistance; for example, an effect of plant growth rate on resistance to barley yellow dwarf virus in barley (Jones & Catherall 1970). There are also a few examples of oligogenic resistance systems involving epistatic effects. In resistance to bean common mosaic virus (BCMV) in *Phaseolus vulgaris*, the *bc-u* locus has no antiviral effect alone, but enhances the antiviral effect of resistance genes at any of three other loci (Drijfhout 1978, Day 1984).

Although Table 1 suggests that resistance to viruses in crop species is in most cases very simple genetically, it should be remembered that the resistant cultivars are often a product of deliberate breeding for this trait. The genetics of virus resistance in wild species might well be more complex, but do not appear to have been investigated. There have been attempts by breeders to construct oligogenic systems of resistance by incorporating several individual resistance genes from related wild species into the commercial cultivar. The best examples are resistance to potato virus Y in potato, and to tobacco mosaic virus (TMV) in tomato.

Models of resistance gene action

A useful way of analysing resistance is to make theoretical models of how resistance genes might work, and of the corresponding response of the virus. Predictions are made from these models and then tested against observations. Fig. 1 shows three types of model. In the positive model (1), the resistant plant produces an inhibitor of the viral replicative cycle, whereas in the negative model (2), the resistant plant lacks some susceptibility function normally required by the virus for pathogenesis. Both models involve some form of recognition event between host- and virus-coded functions which determine the outcome of the interaction in a 'go/no-go' manner. In the positive model, recognition switches on resistance, while in the negative model it switches on susceptibility. The third model is intermediate, in that it does not involve an all-or-nothing response like the first and second, but has a

1. **Positive model: resistance is switched on**

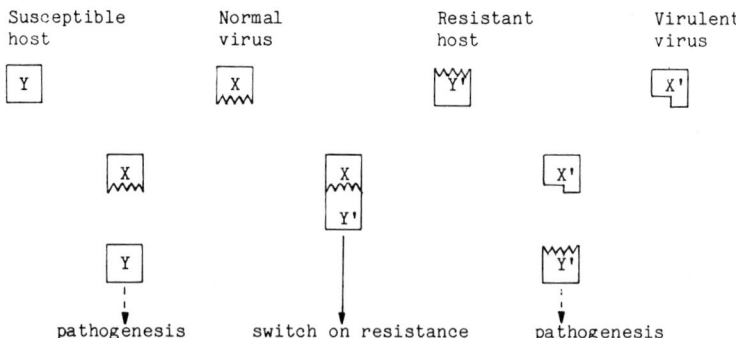

2. **Negative model: susceptibility is not switched on**

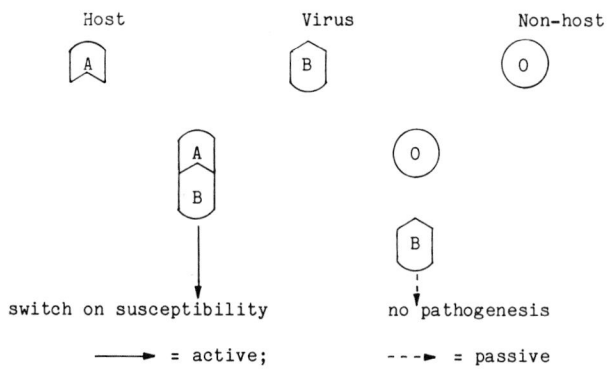

⟶ = active;   ---► = passive

3. **The quantitative interaction model**

molecules specified by:

Normal virus	Virulent virus	Susceptible host	Resistant host
V	V'	H	H'

[V]	+	[H]	⇌ [VH]	susceptibility
[V]	+	[H']	⇌ [VH']	resistance
[V']	+	[H]	⇌ [V'H]	susceptibility
[V']	+	[H']	⇌ [V'H']	susceptibility

FIG. 1.   Three models for interactions between host- and virus-specified molecules which may determine susceptibility or resistance.

response determined by quantitative interactions between host- and virus-specified functions. The outcome, which could be either resistance or susceptibility, depends on the concentration and nature of the recognition product formed. This in turn depends on the concentration and nature of the functions specified by the two participants in the interaction, and can be described by the mathematics of chemical reaction kinetics. The third model could involve either positive or negative mechanisms. Table 2 summarizes some predictions made by the models.

Comparison of models and observations

If we consider only the monogenic resistances, Table 1 shows that most are dominant, with smaller proportions being incompletely dominant or recessive. In fact, the proportion of incompletely dominant alleles is probably an underestimate. Resistance may appear completely recessive or completely dominant when assessed only by scoring visible symptoms, but can show clear

TABLE 2   Predictions from the resistance models

Positive models
(Resistance = inhibition)

Resistance is:
- dominant if recognition is a go/no-go event
- gene-dosage dependent for quantitative interactions
- never fully recessive
- possibly temperature sensitive

Virulence is:
- when a virus function has an altered interaction with the host resistance gene function, or fails to interact

Negative models
(Resistance = reduction or absence of susceptibility)

Resistance is:
- probably recessive for a go/no-go recognition event
- gene-dosage dependent for qualitative interactions
- never fully dominant
- unlikely to be temperature sensitive

Virulence is:
- the ability of the virus to multiply without the host susceptibility factor

gene-dosage dependence (incomplete dominance) when virus multiplication is compared in plants with heterozygous and homozygous resistance alleles (e.g. Fraser & Loughlin 1982, Day 1984). Thus completely recessive behaviour of resistance alleles is probably rare, suggesting that the negative model of resistance (Fig. 1 and Table 2) is also rarer than are positive and quantitative interaction models.

Resistance can be expressed phenotypically in various forms. In many instances, there is localization of the virus, normally with the formation of a necrotic lesion around each site of infection. Other types permit at least some systemic spread of virus but reduce multiplication and symptom severity. A few examples cause apparent complete immunity, in that no symptoms or multiplication are detectable. Table 1 shows that localization is strongly associated with dominant alleles; this is consistent with the operation of a positive mechanism and a go/no-go recognition event. None of the incompletely dominant alleles gives complete localization; some give partial localization, but most allow some systemic spread of virus. This is consistent with the operation of a quantitative interaction model. For the recessive resistances, one example involves localization, but this is anomalous in that the alternative response is complete immunity. Thus the sample contains no case of recessive resistance involving complete localization. Five of the recessive resistances involve apparently complete immunity; this would be consistent with a negative mechanism. Lack of a susceptibility function would be expected to be completely recessive, and to allow absolutely no virus replication.

Temperature sensitivity has not been examined for all resistance genes; those cases where the literature is silent or unclear are listed in Table 1 as 'not known'. For cases where data are available it is clear that temperature sensitivity is strongly associated with dominant alleles, and indeed with those causing localization. This is consistent with the predictions of model 1. Furthermore, it suggests that there may be some feature common to the mechanism of localization in different host species which makes it intrinsically temperature sensitive. Available data suggest that the temperature-dependent transition from localization to systemic spread of virus is rather abrupt. This may imply a heat-induced change of physical state, perhaps in a membrane, rather than simple thermal denaturation of an active inhibitor of virus replication.

Temperature sensitivity has rarely been reported for incompletely dominant alleles. The single case listed in Table 1, resistance to TMV in tomato controlled by the *Tm–1* gene, is peculiar in that it is temperature sensitive for the inhibition of virus multiplication, but temperature resistant for the prevention of symptom formation (Fraser & Loughlin 1982). The two cases of temperature-sensitive resistant alleles might seem to contradict the predictions of the negative model, as there should be no resistance product to be

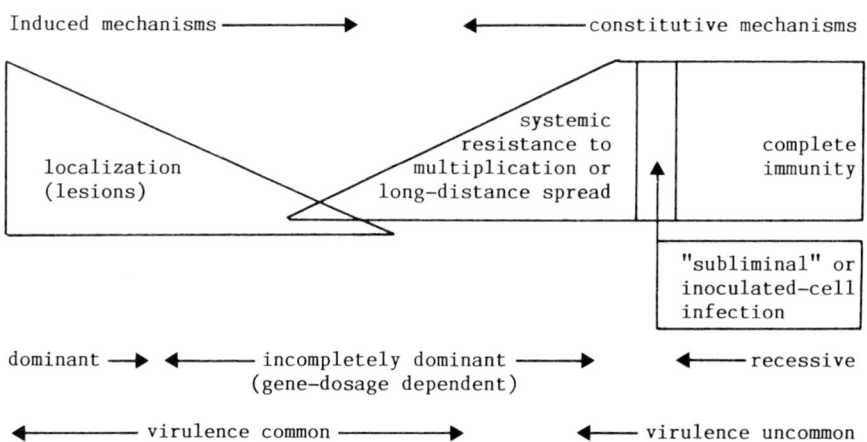

FIG. 2. Some general conclusions about mechanisms of resistance to viruses and the nature of the genetic controls.

temperature sensitive. However, both can be discounted. One is the anomalous case referred to above, and the other appears to involve selection of resistance-breaking 'thermal' strains of the virus.

Fig. 2 summarizes some features of resistance, expressed as a spectrum of responses. The 'subliminal' infections have been reported in a few hosts and appear to allow virus multiplication only in the directly inoculated cells (Sulzinski & Zaitlin 1982). They might be regarded as extreme examples of localization and placed on the left in Fig. 2, but an alternative explanation is that they represent lack of a host function required for cell-to-cell spread.

In so far as data are available, it appears that mechanisms to the left of Fig. 2 are induced; the early stages of virus multiplication in hosts with localizing resistance are similar to those in susceptible hosts (e.g. Takahashi 1973). In contrast, mechanisms to the right of the spectrum may be constitutive. The gene dosage-dependent resistance to TMV controlled by *Tm–1* in tomato operates from the time of inoculation (Fraser & Loughlin 1980). Completely recessive mechanisms involving lack of a susceptibility function are, by definition, constitutive.

Several cases of resistance have now been investigated by inoculation of isolated protoplasts. Generally, it appears that resistance mechanisms involving localization at the level of the intact plant do not operate in protoplasts (e.g. Otsuki et al 1972), whereas those which are systemically effective or cause apparent immunity do operate in protoplasts (e.g. Motoyoshi & Oshima 1977, Barker & Harrison 1984).

Virulence

Fig. 2 suggests that virulent isolates of viruses are more frequent with localization mechanisms than with those tending towards immunity, and more detailed information is given in Table 1. A problem is that for many resistance genes, the literature does not mention whether virulent isolates have been recorded or not. It is also probable that many resistance genes have not been tested against a number of different isolates of the particular virus. These doubtful cases are therefore listed under 'not known'.

Of the remainder, it is clear that very few cases of resistance have proved to be outstandingly durable. The N gene for TMV resistance in tobacco, first described by Holmes (1938), was long thought to be an example of a gene that had not been overcome by a virulent isolate. However, Csillery et al (1983) recently described an isolate from *Capsicum* which effectively overcomes the resistance. Other resistances have been much more ephemeral. Thus the *Tm-1* gene for resistance to the same virus in tomato was widely overcome by virulent isolates within a year of its introduction in commercial cultivars (Pelham et al 1970).

Although the data have limitations because of the large number of 'not known' cases, it appears that virulence is commoner against the dominant and gene dosage-dependent alleles than against apparently recessive alleles. This agrees with the predictions of the models. It may be more difficult for the virus to evolve the ability to replicate without some normally used host function (negative models) than to change so as to fail to interact with a host-specified inhibitor (positive models).

The ability of viruses to evolve virulence, sometimes against more than one host resistance gene, has led to development of gene-for-gene interactions between resistance and virulence in the host and virus. The most complex established to date is the interaction between BCMV and *Phaseolus vulgaris* (Drijfhout 1978).

Conclusion

Although breeding for resistance has given useful control of many economically important viruses, some limitations emerge from this survey. For most crops, the genetic base is very narrow. The high frequency of virulence is a further threat. Indeed, in some cases resistance has been so ephemeral that the genes were hardly worth breeding for. Sometimes no source of resistance is available; J.A. Tomlinson (unpublished work) has listed 22 vegetable crops in the UK alone that lack any resistance to a total of 25 viruses.

These considerations suggest that there is a need to expand the genetic basis of control. This may be achieved by creating oligogenic resistances which are more difficult for the virus to overcome. This aim will become easier to achieve as further types of 'resistance' gene are developed from viral satellites, antisense RNAs and cloned coat protein genes, and by using DNA transformation systems as an alternative to classical breeding.

References

Bald JG, Tinsley TW 1967 A quasi-genetic model for plant virus host ranges. II. Differentiation between host ranges. Virology 32:321–327

Barker H, Harrison BD 1984 Expression of genes for resistance to potato virus Y in potato plants and protoplasts. Ann Appl Biol 105:539–545

Csillery G, Tobias I, Rusko J 1983 A new pepper strain of tomato mosaic virus. Acta Phytopathol Acad Sci Hung 18:195–200

Day KL 1984 Resistance to bean common mosaic virus in *Phaseolus vulgaris* L. PhD Thesis, University of Birmingham, UK

Drijfhout E 1978 Genetic interaction between *Phaseolus vulgaris* and bean common mosaic virus with implications for strain identification and breeding for resistance. Agric Res Rep (Wageningen) 872:1–98

Fraser RSS 1986 Genes for resistance to plant viruses. CRC Crit Rev Plant Sci 3: 257–294

Fraser RSS 1987 Biochemistry of virus-infected plants. Research Studies Press, Wiley, Chichester

Fraser RSS, Loughlin SAR 1980 Resistance to tobacco mosaic virus in tomato: effects of the *Tm–1* gene on virus multiplication. J Gen Virol 48:87–96

Fraser RSS, Loughlin SAR 1982 Effects of temperature on the *Tm–1* gene for resistance to tobacco mosaic virus in tomato. Physiol Plant Pathol 20:109–117

Holmes FO 1938 Inheritance of resistance to tobacco mosaic disease in tobacco. Phytopathology 28:553–561

Holmes FO 1955 Additive resistance to specific virus diseases in plants. Ann Appl Biol 42:129–139

Huber R, Hontilez J, van Kammen A 1981 Infection of cowpea protoplasts with both the common strain and the cowpea strain of TMV. J Gen Virol 55:241–245

Jones AT, Catherall PL 1970 The relationship between growth rate and the expression of tolerance to barley yellow dwarf virus in barley. Ann Appl Biol 65:137–145

Motoyoshi F, Oshima N 1977 Expression of genetically controlled resistance to tobacco mosaic virus infection in isolated tomato leaf mesophyll protoplasts. J Gen Virol 34:499–506

Otsuki Y, Shimomura T, Takebe I 1972 Tobacco mosaic virus multiplication and expression of the N gene in necrotic responding tobacco varieties. Virology 50:45–50

Pelham J, Fletcher JT, Hawkins JH 1970 The establishment of a new strain of tobacco mosaic virus resulting from the use of resistant varieties of tomato. Ann Appl Biol 65:293–297

Sulzinski MA, Zaitlin M 1982 Tobacco mosaic virus replication in resistant and susceptible plants: in some resistant species virus is confined to a small number of initially infected cells. Virology 121:12–19

Takahashi T 1973 Studies on viral pathogenesis in plant hosts. IV. Comparison of early processes of tobacco mosaic virus infection in the leaves of 'Samsun NN' and 'Samsun' tobacco plants. Phytopathol Z 77:157–168

DISCUSSION

Harrison: Dr Fraser, you have described different types of resistance that are controlled by different genes in the plant. Are there any examples of the same type of resistance in a particular plant being controlled by non-allelic genes?

Fraser: Until we have more detailed knowledge of the mechanisms it is hard to give a firm answer. One could argue that in resistance to bean common mosaic virus in *Phaseolus vulgaris*, for example, there are similarities between genes at three separate loci. Unfortunately, we don't generally have a large enough number of resistance genes in a particular species to make that comparison.

Harrison: Is the gene-for-gene idea just a convenient pigeon-hole system? Is it in fact an over-simplification of the actual state of affairs?

Fraser: It is an over-simplification, because it leaves out possible effects of modifier genes or additive effects of different resistance genes at different loci. It also makes some simplistic assumptions about the genetics of virulence in the virus, because here we are not really talking about a gene with the sole function of controlling virulence. It is presumably a gene which is doing something else for the virus and also has this pleiotropic effect. However, the gene-for-gene analysis is useful in that it allows one to make predictions and to see whether the different virus strains or different hosts satisfy these predictions.

Beachy: How many examples are there of genetic traits where there is resistance to viruses from several different virus groups with a given resistance genotype?

Fraser: There are a few examples of an indirect form of resistance effective against several unrelated viruses, such as the avoidance of infection because the plant surface is toughened.

Beachy: What about the type of resistance in which localization occurs?

Fraser: My survey of 63 cases includes one example where a single locus appears to control resistance against two potyviruses, but these are now thought to be closely related anyway.

Dodds: How many of the resistance genes that you have surveyed fail to be as active when an unrelated agent is involved in the biological interaction—for example, an unrelated virus infection? I believe there are cases where a plant that was deemed to be resistant was found to be totally susceptible when it was subject to some other stress, including a second virus infection.

Fraser: There are cases of complementation, or 'assistance', where an unrelated virus can assist an avirulent virus to overcome a resistance mechanism. Perhaps Dr Harrison can comment on this, because of his recent work with potato leafroll virus.

Harrison: Potato leafroll virus (PLRV) is normally confined to phloem tissue in potato or *Nicotiana clevelandii* . However, if one infects *N. clevelandii* with both PLRV and potato virus Y, or a variety of other viruses, PLRV invades some cells which are neither companion cells nor sieve tubes (Barker 1987).

Dodds: In that case you are dealing with a susceptible variety from the start, whereas I was referring to loss of resistance.

Harrison: That is right. We are dealing with a breakdown of a tissue restriction in a susceptible host plant. We also have potato genotypes that are resistant to PLRV, but we have yet to explore whether the same effect occurs in them. Perhaps the Russian work which shows that the *Tm-2* gene in tomato is not effective when the plants are infected with potato virus X is a more relevant example (Taliansky et al 1982).

Dodds: Yes; Dr R.I. Hamilton and I also did that with tobacco mosaic virus (TMV) in cereals (Dodds & Hamilton 1972, Hamilton & Nichols 1977). We found that if barley plants are infected with a cereal virus (either barley stripe mosaic virus (BSMV) or brome mosaic virus), the infected plants are totally susceptible to TMV, with the same type of susceptibility that is found in a susceptible tobacco variety: the virus reached concentrations of 5 mg/g tissue in certain systemically infected leaves. In single infections in barley, TMV belongs to the subliminal class indicated in Fig. 2, with only local and limited replication.

Beachy: Cannot those observations be related, perhaps in the context of complementation of movement function?

Dodds: Yes, it could be related, if you think that a cereal virus such as BSMV can provide a movement function for an unrelated non-cereal virus such as TMV.

Fraser: There might be two alternative types of mechanism. One is a complementation where the resident virus is assisting the guest virus and providing some function that it lacks; the second is a negative mechanism, where the resident virus might be damaging the host and permitting the guest virus to run rampant.

Rybicki: How much of the testing for virus infection in breeding experiments is done by the most sensitive techniques available? As far as I am aware, most of the resistance assessment in cereals, for example, is done on the basis of visible symptoms. Work from our laboratory (Erasmus & von Wechmar 1983a, von Wechmar et al 1984, Rybicki 1984) has shown that rust spore, aphid and seed transmission of brome mosaic virus (BMV) can produce asymptomatic infections of wheat (*Triticum aestivum*). Such BMV infections can lead to a drastic reduction in susceptibility of the plants to stemrust (*Puccinia graminis* f.sp. *tritici;* Erasmus & von Wechmar 1983b). This means that breeding for resistance to the rusts could involve inadvertent breeding for susceptibility to BMV. In addition, the effects of BMV infections alone on wheat are not being properly assessed because breeders are not recognizing the infections.

Fraser: In the past, much of the testing has been done by visual scoring of symptom severity. There is increasing use of techniques such as enzyme-linked immunosorbent assays (ELISA) or cDNA probing. This is useful where the expression of disease is not a clear-cut response, where one is aiming for a quantitative response, and is therefore looking for breeding lines having low rather than moderate or high levels of virus. Also, in any kind of screening or breeding programme of this type one should use as wide a range of virus isolates as possible to be sure that breeding is for a resistance which is likely to be durable. There have been examples of breeding for resistance where screening was performed using only one isolate of the virus; the resistance turned out to be ephemeral in the field.

Bol: To what extent are resistance genes effective at the level of protoplast infection?

Fraser: Referring to my Fig. 2, which shows a spectrum from localization on the left through to immunity on the right, it appears that genes that operate by localizing virus do not operate in protoplasts, whereas those that operate by allowing some systemic spread of virus may do so. However, the number of cases examined is small, so it is a precarious generalization. As examples, the *Tm-1* gene in tomato operates in protoplasts, whereas we know that the *N* gene in tobacco does not.

Zaitlin: You mentioned the category of 'non-host' resistance, Dr Fraser. Normally plants are not susceptible to viruses; resistance is the rule, susceptibility the exception. As an extension to what Dr Bol has just asked, I would like to draw attention to the list of plants that Cheo & Garard (1971) examined with TMV. In almost all species that they investigated they found some limited replication of TMV in the initially infected tissue. How widespread do you think this type of non-host resistance is? Is anyone looking at it?

Fraser: That work is being complemented by experiments looking at virus multiplication in protoplasts of so-called non-hosts. It is too much of a generalization to say that almost any protoplast can replicate almost any virus, but there are cases where protoplasts can replicate viruses for which the plant is a 'non-host'. However, we are in an artificial situation here, because the plants would never normally see virus delivered in that manner to protoplasts. That worries me about how realistic these experiments are, and what they tell us about mechanisms of non-host immunity at the whole plant level.

Zaitlin: You don't have to do it that way. Drs Cheo and Garard tried to get back more virus from the leaf than they put on. In plants that were considered to be subliminally infected they obtained a little more virus than was applied. Therefore this type of resistance doesn't have to be a consequence of being able to infect protoplasts.

Fraser: The distinction to be made is which of the three suggested models of non-host immunity is most probable. The results obtained by Dr Cheo suggest that the negative model (the lack of susceptibility model) is not very likely in these cases. They suggest that we may be dealing with a positive model, and probably with one which acts against cell-to-cell spread after the initial infection event. We also have evidence for what I think of as the 'trivial' model, where, in nature, some kind of physical barrier, such as the cell wall or the epidermis, is the major reason why these non-host plants never see these viruses.

Zaitlin: We don't have good evidence yet on which one applies.

Fraser: No, not until we understand the mechanisms.

Sela: I think Dr Zaitlin was asking whether this interaction is limited by the movement of the virus?

Zaitlin: That is the interpretation. The virus replicates in the initially infected cell but it cannot go anywhere, so in effect you have a resistant host plant.

Sela: If I may extend this, we know of viruses that contribute to their own movement by having genes which enable them to spread. What do we know about host plant functions in this respect? Do you have movement of virus from cell to cell?

Zaitlin: We have no information here. Dr Fraser, you define resistance as the inhibition of virus replication or of symptoms. This is from the standpoint of a plant breeder or horticulturist, I presume, because when a virus is symptomless but replicates substantially that is not really an example of resistance, is it?

Fraser: I was casting my net widely, to include all possible mechanisms. We know of viruses, in tobacco and *Capsicum*, for example, which cause mild symptoms or are cryptic, where there is resistance to symptom formation. However, in many of these cases the amount of virus that is made has not been assessed accurately. Generally, viruses which are cryptic or cause mild symptoms tend not to multiply very well either (Fraser et al 1986). On the other hand, we know of resistance genes which cause complete suppression of symptoms, in terms of visible mosaic formation, but the plant yield can still be dramatically reduced.

Harrison: Mild symptoms do not necessarily indicate a suppression of virus replication. Holmes' masked strain of TMV multiplies strongly in tobacco, yet one can hardly tell that the plants are infected.

Fraser: So far as we have been able to study that with the isolate we have today, we find that the masked strain doesn't multiply as well as the wild-type.

van Vloten-Doting: Does your list of crop species that show resistance to viruses include crops carrying transposable elements? If so, you could try to select for plant mutants that have lost their resistance. That would confirm your hypothesis that resistance is a positive response.

Fraser: There are many examples of resistance genes in maize which would be worth looking at from that point of view.

van Vloten-Doting: Is anybody investigating this?

Fraser: There is some evidence that viruses may interact with transposable elements and for a possible activation of transposon activity after virus infection, but I do not know of any cases yet where transposable elements have eliminated resistance. An interesting approach to the isolation of resistance genes would be to look for plants with a dominant resistance gene in the heterozygous state where resistance had failed phenotypically.

van Kammen: In Table 1 you mentioned 50 host–virus combinations where resistance was monogenic. Of these 50 combinations, how many concern the same host plant and a different virus? And how different are the resistance genes in these combinations?

Fraser: There is a fairly wide range of hosts (about 20), but several appear a few times—tomato, potato, *Phaseolus vulgaris*, tobacco, cucumber. I haven't been able to see any strong common feature in particular hosts.

van Kammen: You don't find, say, four or six genes for resistance against RNA viruses in cucumber?

Fraser: One could make a statement like that for a particular plant species but I'm not sure that you could say anything further about common mechanisms.

Harrison: I suspect that some of the categories which you defined tend to blur at the edges, Dr Fraser! You mentioned subliminal infections. A specific example might be the extreme resistance to potato virus X seen in potato seedling 41956. Although some people consider this seedling to be immune, it is possible, with effort, to find evidence that the virus induces a necrotic reaction in a very few cells. This example of resistance is controlled by a dominant gene, so it doesn't fit in very well with the pattern you outlined.

Fraser: It is inevitable, because we are dealing with a wide range of biological material, that our definitions will blur at the edges, as a reflection of the variability of the responses that we are investigating. As I said, I am prepared to see the subliminal infections being moved from the right-hand side of Fig. 2 to the left-hand side. Perhaps we even have to put them in both positions!

Harrison: From the practical point of view, the important point is the ease with which the resistance is liable to be broken in field conditions. You are suggesting that one type of single-gene resistance is more likely to break down than another.

Fraser: Yes, the prediction from the model is that negative, recessive resistance is less liable to be broken. But we have insufficient data for the apparently recessive resistance genes to test that properly. Out of 13 cases that I examined, eight or nine were 'don't knows' in that they had apparently not been fully tested. High durability has been proved in a few cases.

If a gene is dominant, I would argue that it cannot be operating by a negative (lack of susceptibility) mechanism, and therefore that it should not fulfil the prediction of virulence being a difficult attribute for the virus to acquire. The possibility that negative mechanisms are difficult to overcome applies only to fully recessive genes. My sample contains five possible examples, out of 63, of fully recessive genes which could involve negative mechanisms. Therefore, if negative resistance does exist, it is a minority case.

Bol: It is difficult to analyse the structure of plant resistance genes, but it is possible to analyse the genome of a virus isolate that has succeeded in overcoming resistance, or has at least become an attenuated strain. We can make chimeric molecules at the DNA level from the genomes of virulent and attenuated strains and look for the virus gene that is involved in pathogenicity. If you compare replicase genes, transport genes, or structural genes of the virus, which is most likely to be involved in the host response?

Fraser: To answer that obliquely, the most interesting and promising way to study the function of resistance genes is to compare virus isolates which differ in their response. The high mutation frequency of virus genomes is a problem, and it will be hard to pin down a particular change in base sequence to the acquisition of virulence, because of the background noise; but that kind of information is increasingly available and will provide an interesting way forward. From the examination of pseudorecombinants between viruses with multi-component genomes it does seem that virulence determinants can be scattered across different components of the genome. We have to consider different kinds of virulence as being based on different viral functions in different host–virus combinations.

van Vloten-Doting: We have sequenced a mutant of alfalfa mosaic virus (AlMV) which, in contrast to the parent stain, infects bean plants systemically. The mutant characteristic maps on RNA-2. There are over 60 mutations in the mutant RNA-2, so it is very difficult to say which mutation is responsible for the difference in phenotype.

Fraser: Is that a natural mutant or an artificial one?

van Vloten-Doting: It was derived from a programme in which we used UV irradiation of the middle component (the one carrying RNA-2). The problem with this sort of study is that you have both induced and spontaneous mutations. I would guess that the original mutation was due to UV irradiation. Subsequently the mutant apparently accumulated many mutations. During selection of this mutant we saw changes in phenotype. The first mutation probably reduced the 'fitness' (ability to replicate) of the virus. Any spontaneous mutation which increases the fitness may accumulate until a new sequence with optimal fitness has emerged. This mutant is not far removed genetically from the parent strain, because it has been replicated for a few months (although we do not know how many generations that involves); nevertheless it carries an enormous number of mutations.

Harrison: That is an interesting example. I hope we shall explore some of the effects of mutations on the biological fitness of viruses later in the symposium, particularly in relation to virulence.

References

Barker H 1987 Invasion of non-phloem tissue in *Nicotiana clevelandii* by potato leafroll virus is enhanced in plants also infected with potato Y potyvirus. J Gen Virol 68:1223–1227

Cheo PC, Garard JS 1971 Differences in virus-replicating capacity among plant species inoculated with tobacco mosaic virus. Phytopathology 61:1010–1012

Dodds JA, Hamilton RI 1972 The influence of barley stripe mosaic virus on the replication of tobacco mosaic virus in *Hordeum vulgare* L. Virology 59:418–427

Erasmus DS, von Wechmar MB 1983a The association of brome mosaic virus and wheat rusts: I. Transmission of BMV by uredospores of wheat stem and leaf rust. Phytopathol Z 108:26–33

Erasmus DS, von Wechmar MB 1983b Reduction of susceptibility of wheat to stem rust (*Puccinia graminis* f.sp. *tritici*) by brome mosaic virus. Plant Dis 67:1196–1198

Fraser RSS, Gerwitz A, Morris GEL 1986 Multiple regression analysis of the relationships between tobacco mosaic virus multiplication, the severity of mosaic symptoms, and the growth of tobacco and tomato. Physiol Mol Plant Pathol 29:239–249

Hamilton RI, Nichols C 1977 The influence of bromegrass mosaic virus on the replication of tobacco mosaic virus in *Hordeum vulgare*. Phytopathology 67:484–489

Rybicki EP 1984 Investigations of viruses affecting South African small grains. PhD Thesis, University of Cape Town

Taliansky ME, Malyshenko SI, Pshennikova ES, Atabekov JG 1982 Plant virus-specific transport function. II. A factor controlling virus host range. Virology 122:327–331

von Wechmar MB, Kaufman A, Desmarais F, Rybicki EP 1984 Detection of seed-transmitted brome mosaic virus by ELISA, radial immunodiffusion and immunoelectroblotting tests. Phytopathol Z 109:341–352

Resistance of cowpeas to cowpea mosaic virus and to tobacco ringspot virus

George Bruening, Fernando Ponz, Christopher Glascock, Mary L. Russell, Adib Rowhani* and Catherine Chay

*Department of Plant Pathology and *Agricultural Research Service of the US Department of Agriculture, College of Agricultural and Environmental Sciences, University of California, Davis, California 95616, USA*

Abstract. Cowpea mosaic virus (CPMV) and tobacco ringspot virus (TobRV), though members of distinct virus groups, each have two genomic RNAs separately encapsidated and generate functional proteins by the specific proteolysis of polyproteins. CPMV and TobRV infect most cowpea (*Vigna unguiculata*) lines, including line Blackeye 5. Cowpeas of line Arlington did not support a detectable increase of CPMV, as previously reported, or of TobRV. In F_2 progeny of cowpea crosses the resistances to CPMV and to TobRV were inherited as distinct, simple dominant traits. Extracts of Arlington cowpea leaves have three activities that are candidate mediators of resistance to CPMV: proteinase(s) that degrade CPMV proteins, inhibitor(s) of the translation of CPMV RNAs, and an inhibitor of proteolytic processing of the polyprotein precursor to the coat proteins. These activities were tested for their virus specificity and inheritance in the progeny of cowpea crosses. The proteinase inhibitor showed the specificity and co-inheritance expected for a mediator of resistance to CPMV. No inhibitor of the processing of TobRV polyproteins was detected in similar extracts, indicating that the resistances to CPMV and to TobRV may be controlled by different mechanisms as well as by different genes.

1987 Plant resistance to viruses. Wiley, Chichester (Ciba Foundation Symposium 133) p 23–37

Cowpea protoplasts resistant to CPMV-SB

Some years ago researchers in this laboratory accidentally exhausted our supply of seeds of cultivar Blackeye 5 cowpea (*Vigna unguiculata*), a host for the SB isolate of cowpea mosaic virus (CPMV-SB). They turned to another cowpea line and discovered, to their surprise, that it failed to support CPMV-SB. This result stimulated a survey of more than 1000 cowpea lines, of which 65 proved to be immune to CPMV-SB according to specific criteria (Beier et al 1977): seedlings must remain free of symptoms and detectable virus or capsid antigen after being inoculated with a 100 times greater concentration of CPMV-SB than that which uniformly infected susceptible cowpeas.

The 'operational immunity' of seedlings from lines that satisfied the above criteria has proved to be very reliable. Immune cowpeas did not become infected when grafted to CPMV-SB-infected Blackeye 5 cowpeas, nor was the immunity overcome by environmental or hormonal stresses (Beier et al 1979 and unpublished results). The observed immunity of these seedlings provided an extreme and biochemically useful contrast to susceptibility. However, the protoplasts from the seed leaves of almost every immune cowpea supported an increase in CPMV-SB as effectively as did the protoplasts from susceptible lines.

One line, Arlington, proved to be the one-in-a-thousand cowpea from which protoplasts were recovered that did not support a strong increase in CPMV-SB. Arlington cowpea protoplasts accumulated only about one-tenth the amount of CPMV-SB accumulated by protoplasts of the susceptible line Blackeye 5 and the other immune lines, when protoplasts were inoculated with virions at 1 to 4 µg/ml (Beier et al 1979, Kiefer et al 1984). CPMV-SB is a member of the comovirus group. Another comovirus, cowpea severe mosaic virus strain DG (CPSMV-DG), infects virtually all cowpeas and cowpea protoplasts to which it has been inoculated, including Arlington (Beier et al 1979 and unpublished results). Thus Arlington cowpeas exhibit specific virus resistance.

How Arlington cowpea protoplasts resist CPMV-SB

We have reason to believe that CPMV-SB which has been inoculated to Arlington cowpea leaves reaches the cytoplasm of leaf cells. The number of local lesions induced by CPSMV-DG (Bruening et al 1979) and the yield of CPSMV-DG (Eastwell et al 1983) from inoculated Arlington cowpea primary leaves are both reduced if CPMV-SB is included in the inoculum. This protection against CPSMV-DG is exhibited also by CPMV-SB RNA but not by ultraviolet-inactivated CPMV-SB or by empty capsids of CPMV-SB. Ponz & Bruening (1986) introduced the term 'concurrent protection' to denote this phenomenon, since the protecting virus (CPMV-SB) and the challenging virus (CPSMV-DG) must be co-inoculated for maximum protective effect. We postulate that the inoculated CPMV-SB or CPMV-SB RNA must reach the cytoplasm of Arlington cowpea leaf cells and express at least some virus functions to exert the observed effects.

Results from genetic crosses of Arlington and Blackeye 5 cowpeas indicate that seedling immunity to CPMV-SB is inherited as a simple dominant character. Immune progeny seedlings are sources of resistant protoplasts, reinforcing the suggestion from the concurrent protection experiment that both Arlington cowpea seedling immunity and Arlington cowpea protoplast resistance reflect a restriction of virus replication at the cellular level (Kiefer et al 1984). Inheritance of seedling immunity as a dominant character suggests that the immunity is mediated by an inhibitor of virus replication rather than by the absence of some factor required by the virus.

Kiefer et al (1984) exploited the residual replication of CPMV-SB in Arlington cowpea protoplasts to analyse the accumulation of virions and of intermediates in CPMV-SB replication. Immunofluorescence assays revealed a high level of accumulation of CPMV-SB capsid antigen in Blackeye 5 protoplasts and a generally low level of CPMV-SB in the inoculated Arlington cowpea protoplasts, rather than a few strongly fluorescent cells. This result indicates that the observed low level of CPMV-SB yield from Arlington cowpea protoplasts cannot be ascribed to efficient replication in just a few cells, but rather reflects a general réstriction on CPMV-SB accumulation in all cells.

The results of assays for CPMV-SB $(+)$RNA (i.e. RNA of the encapsidated, messenger polarity), CPMV-SB $(-)$RNA and capsid antigen in inoculated Blackeye 5 and Arlington cowpea protoplasts were explained by Kiefer et al (1984) as the consequences of a restriction by Arlington cowpea protoplasts of the production of CPMV-SB-encoded proteins. Several mechanisms could account for reduced accumulation of CPMV-SB proteins in Arlington cowpea protoplasts, as compared to the accumulation in Blackeye 5 protoplasts. Among them are degradation of CPMV-SB RNA(s), reduced translation of CPMV-SB RNA(s), and degradation of CPMV-SB proteins. A fourth possibility is a restriction of the proteolytic processing of CPMV-SB polyproteins (Pelham 1979, Franssen et al 1984ab), a reaction that is characteristic of the replication of comoviruses and the members of a few other plant virus groups. This has been demonstrated to be a virus-specific process (Goldbach & Krijt 1982).

Comoviruses have two single-stranded RNA chromosomes of the messenger or $(+)$ polarity that are separately encapsidated in icosahedral shells composed of 60 copies each of two coat proteins. The middle component (M) and the more rapidly sedimenting bottom component (B) of a comovirus each contains a single RNA chromosome, RNA-2 and RNA-1, respectively. Each of these virion RNAs bears a 5'-linked protein and 3' polyadenylate. Each encodes a polyprotein from which functional proteins are derived by specific proteolysis.

The two *in vitro* translation products of RNA-2, initiated at two AUGs, have apparent M_r values of 105 000 (105K) and 95 000 (95K). An assay for the cleavage of these polyproteins requires the addition of a source of a CPMV-SB proteinase, either extracts of infected protoplasts or translation products of RNA-1 (Franssen et al 1982, Pelham 1979). The immediate processing products of the more abundant 95K protein are polypeptides of apparent M_r values 48 000 (48K, from the N-terminal portion of 95K) and

60 000 (60K, the precursor of the two capsid polypeptides of CPMV-SB). By applying such an assay, Sanderson et al (1985) detected an activity in extracts of Arlington cowpea protoplasts that inhibited the formation of the 60K and 48K polypeptides. Extracts of Blackeye 5 cowpea protoplasts had negligible amounts of this activity. Thus an inhibitor of the proteolytic processing of an RNA-2-encoded polyprotein was implicated as a, perhaps the only, mediator of resistance in Arlington cowpea protoplasts. These results set the stage for an investigation of the immunity that Arlington cowpea seedlings exhibit against CPMV-SB and of a recently discovered immunity of these seedlings to tobacco ringspot virus (TobRV).

How Arlington cowpea seedlings resist CPMV-SB

The characteristics of the reliable immunity of Arlington cowpeas against CPMV-SB, as described above, suggest the presence of a powerful and specific inhibitor of CPMV-SB replication. Results from analyses of infected protoplasts implied that such an inhibitor should act on one or more of the proteolytic polyprotein-processing reactions that are engendered by CPMV-SB in infected cells. We found that extracts of Arlington cowpea leaves actually exhibited three activities that could conceivably restrict the replication of CPMV-SB. These are activities that are less apparent or undetected in similar extracts from Blackeye 5 cowpeas: protease(s) that degrade CPMV-SB proteins; an activity that interferes with the translation of CPMV-SB RNAs; and the inhibitor(s) of CPMV-SB polyprotein proteolytic processing (Ponz et al 1987a).

To determine whether any, or several, of these activities mediates the immunity of Arlington seedlings to CPMV-SB we required either assays of sufficient specificity, or the ability to separate the activities. We reasoned that an activity of interest should be ineffective against CPSMV-DG and should be inherited in parallel with immunity against CPMV-SB in progeny from crosses of Arlington and Blackeye 5 cowpeas. An extract was prepared by homogenizing cowpea leaves in buffer and centrifuging the suspension. The supernatant was chromatographed on cross-linked dextran beads (Sephadex G-25) to remove low molecular weight contaminants. After concentration the void volume was applied to cross-linked 6% agarose beads (Sepharose CL-6B). Four peaks of protein (I–IV) were eluted. Most of the proteinase(s) were confined to peaks II and III. Peak IV contained the inhibitors of translation and polyprotein processing.

Fig. 1 presents the results from an assay that allowed the inhibitor(s) of polyprotein processing to be assayed even in the presence of the inhibitor(s) of translation. The assay is based on our observation that, during a one to two hour incubation of CPMV-SB RNAs with rabbit reticulocyte lysate, most of the translation occurs during the first 30 minutes and most of the polyprotein

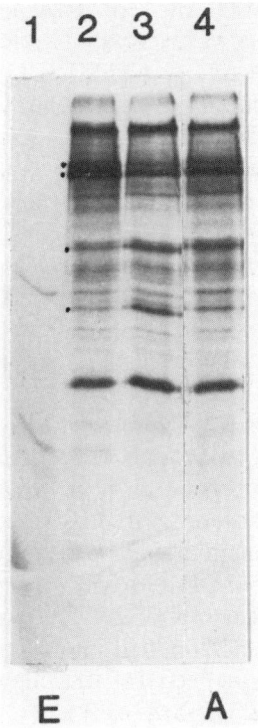

FIG. 1. An activity from extracts of Arlington cowpea seedlings that inhibits the proteolytic processing of a CPMV-SB polyprotein. Lane 1: products of endogenous (E) incorporation, no added RNA. [35S]Met-labelled polypeptides were resolved by electrophoresis through a 12.5% polyacrylamide gel and detected by fluorography. The polyprotein translation products of a mixture of CPMV-SB RNAs 1 and 2 (1 μg/15 μl reaction solution) remained for the most part unprocessed 30 minutes after incubation of the reticulocyte lysate reaction solution was initiated (lane 2). The zones of the 105K and 95K polyprotein translation products of RNA-2 and the 60K and 48K processing products of 95K are located by four dots to the left of lane 2. After 60 minutes, the 60K and 48K products of the processing of 95K accumulated to easily detected levels (lane 3). An extract of Arlington cowpea leaves was fractionated by chromatography on 6% agarose beads. When an aliquot of undiluted solution from peak IV of eluted Arlington (A) cowpea proteins was added to the translation reaction mixture at 30 min and incubation was continued to 90 min, processing was observed to be inhibited (lane 4). The peak IV from an extract of Blackeye 5 cowpea leaves did not interfere with processing (not shown).

processing occurs subsequently. If an inhibitor-containing fraction is added at 30 min its greatest effect will be on polyprotein processing. If the fraction to be tested is added at zero time rather than 30 min, the assay detects inhibitor(s) of translation. Thus, by a combination of fractionation and specific assays, the three activities could be assessed independently.

The three activities were tested for their virus specificity and co-inheritance with the immunity trait in cowpea crosses. Proteinase(s) from extracts of Arlington cowpea leaves degraded CPSMV-DG and CPMV-SB proteins equally well (Ponz et al 1987a). One homozygous susceptible F_3 progeny line produced a stronger proteinase activity against CPMV-SB proteins than was observed for an immune progeny line, indicating lack of co-inheritance with immunity. The inhibitor(s) of *in vitro* translation were up to 10-fold more potent against *in vitro* translation of CPMV-SB RNA than of CPSMV-DG RNA. Extracts from homozygous immune progeny of a seven step backcross series inhibited translation of CPMV-SB RNAs more effectively than extracts from the corresponding susceptible progeny. Thus one or more inhibitors of translation, but no detected proteinase, may be contributing to the immunity of Arlington cowpeas against CPMV-SB.

In contrast to the activities that degrade CPMV-SB proteins and interfere with the translation of CPMV-SB RNAs, the inhibitor(s) of polyprotein processing were not effective in an assay in which translation products of CPSMV-DG RNA translation were tested. Dilutions of the peak IV fraction from Sepharose CL-6B chromatography, which prevented the generation of 60K and 48K proteins from CPMV-SB 95K polyprotein, had no detectable effect on the *in vitro* processing of CPSMV-DG polyproteins. The activity in extracts of Arlington cowpea leaves that interferes with polyprotein processing has the virus specificity expected if its action in fact causes Arlington cowpeas to be immune to CPMV-SB.

The Arlington cowpea was backcrossed to Blackeye 5 seven times, selection at each cross being only for the trait of immunity to CPMV-SB. From the progeny of the seventh backcross, which should be nearly isogenic with Blackeye 5, we generated, by selfing and selection, lines that are homozygous for immunity and lines that are homozygous for susceptibility. The peak IV fraction from extract of leaves of a homozygous immune cowpea prevented the processing of the 95K polypeptide; the corresponding fraction from a homozygous susceptible cowpea, like the fraction from Blackeye 5 cowpea, did not. Ponz et al (1987a) conclude that the immunity of Arlington cowpeas against CPMV-SB is probably mediated by an inhibitor of one or more of the processing reactions by which CPMV-SB polyproteins are cleaved to generate functional proteins. Proteinase inhibitors that are proteins are well-known constituents of plants. The inheritance of immunity as a simple dominant trait suggests that the product of the 'immunity gene' and the inhibitor may be the same entity, a protein.

Arlington cowpeas are immune to tobacco ringspot virus

In a comparison of the reactions of Arlington and Blackeye 5 cowpeas to various viruses, the former was found to be immune to TobRV (Ponz et al

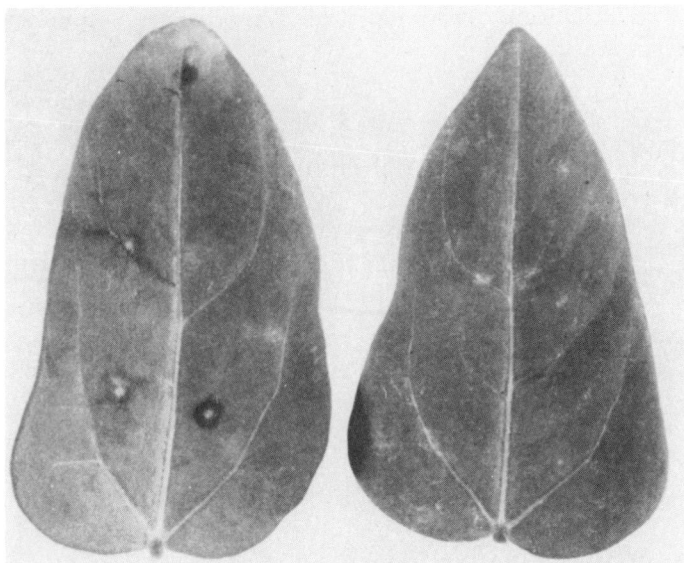

FIG. 2. Comparison of the reaction of Arlington cowpea seed leaves to inoculation with 6 μg/ml CPSMV–DG (left leaf) and 40 μg/ml TobRV. A low concentration of CPSMV–DG was necessary in order to prevent early collapse of the inoculated leaves. The photograph was taken five days after inoculation. CPSMV–DG induced characteristic ringed necrotic lesions, whereas TobRV induced no symptoms.

1987b) by criteria similar to those applied in the earlier survey of cowpea lines (Beier et al 1977). This observation is of special interest because of the similarities between TobRV, a member of the nepovirus group, and the comoviruses. Both nepoviruses and comoviruses have two chromosomal, single-stranded RNAs bearing 5′-linked protein and 3′ polyadenylate. Genes are expressed by translation of the RNAs into polyproteins that are proteolytically processed.

Fig. 2 compares the responses of inoculated Arlington cowpea seed leaves to TobRV and CPSMV-DG. In contrast, TobRV-inoculated seedlings of the cultivar Blackeye 5 developed firstly large necrotic local lesions on the inoculated leaves, then stem necrosis. The plants collapsed and died, usually before the development of the trifoliate leaves. Enzyme-linked immunosorbent assays (ELISA) (Rowhani et al 1985) for TobRV capsid antigen revealed that TobRV- and buffer-inoculated Arlington cowpea seed leaves and subsequently developing trifoliate leaves all gave comparable background ELISA values. Under similar conditions, extracts from inoculated leaves of Blackeye 5 cowpeas gave off-scale ELISA values unless the extract was diluted.

F_1 progeny cowpeas of a cross of Arlington to Blackeye 5 were all immune to TobRV. The F_2 progeny segregated 3:1 (immune:susceptible) (Ponz et al 1987b). Thus, like its immunity against CPMV-SB, the immunity of the

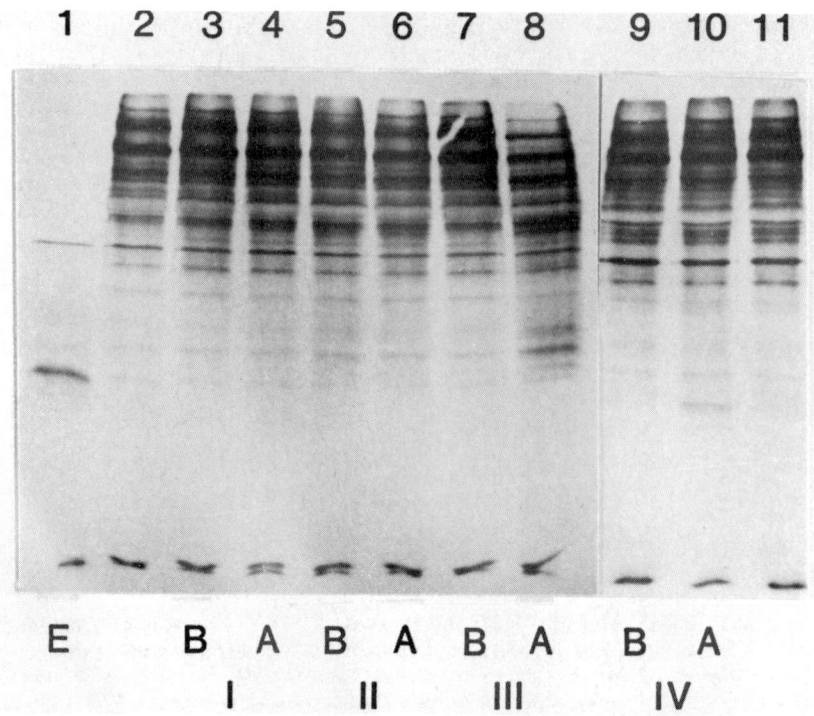

FIG. 3. Extract of Arlington cowpea leaves did not interfere with the proteolytic processing of TobRV polyproteins. Conditions for translation of a mixture of TobRV RNAs 1 and 2 are similar to those for the experiment of Fig. 1, except that additions of extract were at 40 min and the incubation was continued to 120 min. Lane 1: endogenous incorporation (E), no added RNA. Lanes 3, 5, 7 and 9: additions of chromatography fractions I, II, III and IV, respectively, from an extract of Blackeye 5 cowpea leaves (B). Lanes 4, 6, 8 and 10: similar additions from an extract of Arlington cowpea leaves (A). Lanes 2 and 11: addition of buffer only. The first eight and the last three lanes are from different experiments. Note that the proteinase activity of the Arlington cowpea peak III fraction has partially obscured the results. This activity has previously been observed with CPMV–SB and CPSMV–DG proteins as substrate.

Arlington cowpea against TobRV is controlled as if it is a simple dominant trait. To test whether the resistance to the two viruses was controlled by the same gene, F_2 progeny that were found to be immune to TobRV were inoculated with CPMV-SB. A 3:1 ratio (immune:susceptible) was observed. In a reciprocal experiment, F_2 progeny that were immune to CPMV-SB also gave a 3:1 ratio when subsequently inoculated with TobRV. The immunity against TobRV and that against CPMV-SB are controlled by distinct genes.

  Although immunity against the two viruses is controlled by distinct genes, they could have similar mechanisms. The proteolytic processing of TobRV polyproteins *in vitro* has been reported (Forster & Morris-Krsinich 1985,

Jobling & Roger Wood 1985). We observed a similar pattern of processing and noted that, as with CPMV-SB polyprotein processing *in vitro*, little processing of TobRV polyproteins occurred in the first 40 min after the translation of a mixture of RNAs 1 and 2 was initiated in a rabbit reticulocyte lysate. Assays for processing of TobRV polyproteins, based upon this timing, were applied to four fractions of an Arlington cowpea leaf extract (peaks I–IV) eluted from Sepharose CL-6B. The controls were similar fractions from an extract of Blackeye 5 cowpea leaves. The results (Fig. 3) gave no indication of inhibition of polyprotein processing. Of course, this negative result does not eliminate a mechanism based on interference with polyprotein processing (Ponz et al 1987b).

We have summarized here the virus-resisting capabilities of the unusual cowpea line Arlington. This line, the only known source of protoplasts that restrict the replication of CPMV-SB, is also immune to TobRV. However, the immunity of Arlington seedlings against the two viruses is clearly controlled by independent genes, and apparently by different mechanisms.

Acknowledgements

F. Ponz is a Fulbright Scholar from Spain; C.A. Chay is a McKnight Foundation Training Grant Fellow. Research in this laboratory on the mechanisms of resistance of plants to plant viruses has been supported by a grant from the Division of Biological Energy Research, Department of Energy, DE-FG03-85ER13353, by the Biotechnology Research and Education Program of the University of California, and by the Agricultural Experiment Station of the University of California.

References

Beier H, Siler DJ, Russell ML, Bruening G 1977 Survey of susceptibility to cowpea mosaic virus among protoplasts and intact plants from *Vigna sinensis* lines. Phytopathology 67:917–921

Beier H, Bruening G, Russell ML, Tucker CL 1979 Replication of cowpea mosaic virus in protoplasts isolated from immune lines of cowpeas. Virology 95:165–175

Bruening G, Lee S-L, Beier H 1979 Immunity to plant virus infection. In: Sharp WR et al (eds) Plant cell and tissue culture, principles and applications. Ohio University Press, Columbus, p 421–440

Eastwell KC, Kiefer MC, Bruening G 1983 Immunity of cowpeas to cowpea mosaic virus. In: Goldberg RB (ed) Plant molecular biology. UCLA (Univ Calif Los Ang) Symp Mol Cell Biol New Ser 12:201–211

Forster RLS, Morris-Krsinich BAM 1985 Synthesis and processing of the translation products of tobacco ringspot virus in rabbit reticulocyte lysates. Virology 114:516–519

Franssen H, Goldbach R, Broekhuijsen M, Moerman M, van Kammen A 1982 Expression of middle-component RNA of cowpea mosaic virus: in vitro generation of a precursor to both capsid proteins by a bottom-component RNA-encoded protease from infected cells. J Virol 41:8–17

Franssen H, Goldbach R, van Kammen A 1984a Translation of bottom component

RNA of cowpea mosaic virus in reticulocyte lysate: faithful proteolytic processing of the primary translation product. Virus Res 1:39–49

Franssen H, Moerman M, Rezelman G, Goldbach R 1984b Evidence that the 32,000-Dalton protein encoded by bottom-component RNA of cowpea mosaic virus is a proteolytic processing enzyme. J Virol 50:183–190

Goldbach R, Krijt J 1982 Cowpea mosaic virus-encoded proteinase does not recognize primary translation products of mRNAs from other comoviruses. J Virol 43:1151–1154

Jobling SA, Roger Wood K 1985 Translation of tobacco ringspot virus RNA in reticulocyte lysate: proteolytic processing of the primary translation products. J Gen Virol 66:2589–2596

Kiefer MC, Bruening G, Russell ML 1984 RNA and capsid accumulation in cowpea protoplasts that are resistant to cowpea mosaic virus strain SB. Virology 137:71–81

Pelham HRB 1979 Synthesis and processing of cowpea mosaic virus proteins in reticulocyte lysates. Virology 96:463–477

Ponz F, Bruening G 1986 Mechanisms of resistance to plant viruses. Annu Rev Phytopathol 24:355–381

Ponz F, Glascock CB, Bruening G 1987a An inhibitor of polyprotein processing with the characteristics of a natural virus resistance factor. Molec Plant–Microbe Interac, in press

Ponz F, Russell ML, Rowhani A, Bruening G 1987b A cowpea line has distinct genes for resistance to tobacco ringspot virus and to cowpea mosaic virus. In press

Rowhani A, Mircetich SM, Shepherd RJ, Cucuzza JD 1985 Serological detection of cherry leafroll virus in English walnut trees. Phytopathology 75:48–52

Sanderson JL, Bruening G, Russell ML 1985 Possible molecular basis of immunity of cowpeas to cowpea mosaic virus. In: Key JL, Kosuge T (eds) Cellular and molecular biology of plant stress. UCLA (Univ Calif Los Ang) Symp Mol Cell Biol New Ser 22:401–412

DISCUSSION

van Vloten-Doting: Have you tried to isolate cowpea mosaic virus from your Arlington-infected protoplasts and then infect plants again, to see whether the isolated virus replicates to a higher extent?

Bruening: We recover no virus from plants, so that experiment is not possible. As far as we know, the virus that is recovered from the resistant protoplasts is the same in every way as the inoculated virus. It has the same specific infectivity and resistance to ribonuclease.

Bol: If you inoculate the Arlington protoplasts with a mixture of the SB (CPMV) and DG (CPSMV) strains, does the cowpea severe mosaic virus permit replication of the cowpea mosaic virus RNA?

Bruening: Are you asking whether the interference phenomenon holds up when we examine protoplasts? It is hard to do that experiment because, unlike the situation for the whole plant, the CPMV capsid alone, in the form of top component, is effective in protoplasts in interfering with DG. One of our controls in studies of the ability of cowpea mosaic virus to interfere with cowpea severe mosaic virus replication was to test whether RNA from the virus and the top component had the same effect. In the intact plant, if you inoculate top component you do not get protection. If you inoculate the RNA, you get protection. However, inoculation of protoplasts with top component of CPMV is as effective as inoculating virus particles with regard to the ability to interfere with CPSMV accumulation. So we had to conclude that the protoplast system is not a reasonable model for the interference observed in the intact plant, and that no conclusions can be drawn from such an experiment.

Fritig: Are your proteinase inhibitors related to those described by Ryan's group, which are produced upon mechanical injury of many plants (Ryan 1973, Gustafson & Ryan 1976)?

Bruening: I must acknowledge an intellectual debt to Professor Ryan because his work stimulated us to look at proteinase inhibitors. We don't believe that the inhibitor we are studying is like his. Our inhibitor seems to be less heat stable.

Harrison: Is the inhibitor of proteolytic processing present in healthy plants or is its synthesis induced by some kind of stimulus?

Bruening: Our evidence so far is that it is constitutive, but this is far from proven.

Harrison: Do you know anything about its mode of action? Does it bind to the viral polyprotein?

Bruening: We presume that in order to act it must bind, but we have no evidence for that. Plants contain many inhibitors of proteinases. We think that this inhibitor is a protein that is itself an inhibitor of proteinases, specifically a proteinase of CPMV. My student, Christopher Glascock, incubated the Arlington plant extract that contains the proteinase inhibitory activity (peak IV on the Sepharose CL-6B column) with immobilized trypsin. This inactivated the inhibitor of CPMV polyprotein processing. We therefore believe that the inhibitor is itself a protein, perhaps the direct product of what appears to be a simple dominant resistance gene.

Beachy: How far have you been able to purify the proteinase in peak IV? Do you have a fraction from HPLC which contains the enriched activity?

Bruening: We do not yet have one zone on a gel that we could identify. Chris Glascock has made some progress on purification of the inhibitor of polyprotein processing. Our goal is to purify this protein and obtain enough amino acid sequence data to synthesize an oligodeoxyribonucleotide designed to be complementary to the mRNA.

Nishiguchi: Are there any differences in the total amounts of proteinase present in extracts from seedlings and protoplasts?

Bruening: We have not compared extracts from plants and protoplasts directly.

Matthews: Could differential intracellular compartmentation of some of the enzyme activities be confusing the interpretation of your results? Presumably you do a low speed centrifugation and have a pellet. Could artefactual events be occurring there?

Bruening: We cannot claim to have examined every possible source of activity, because of fractionation problems. It is likely that some of the activities, such as the general proteinase activity, are vacuolar or reside between cells. Perhaps this is why these activities did not trouble us when we were working with protoplasts, but troubled us a great deal when we began to work with the intact plant.

Hohn: Is the polyprotein proteinase of the virus virus-coded or host-coded? Also, did you test proteinase inhibitors such as eglin and Ryan's proteinase inhibitor *in vitro* ?

Bruening: We have not really tested other proteinase inhibitors. The research groups of van Kammen and Goldbach have found that the proteinase that processes polyproteins is virus-encoded (Goldbach & Krijt 1982, Franssen et al 1984).

Goldbach: We have recently found that there is only one proteinase specified by CPMV—the 24K protein encoded by the B RNA. Formerly we thought that the 32K protein, also encoded by the B RNA, represented a second proteinase recognizing the Q/M cleavage sites in the viral polyproteins. Recent experiments show that this 32K protein is a cofactor needed for the 24K proteinase to cleave the Q/M cleavage site in the M polyprotein (i.e. in *trans*). So if it is correct that you have now detected an inhibitor of this Q/M cleavage, it could be either a proteinase inhibitor, or an inhibitor of the 32K cofactor.

You assayed for inhibition of the Q/M cleavage in the M polyprotein; is the Q/M cleavage in the B polyprotein (for which the presence of 32K is not necessary) also inhibited? Additionally, did you test whether B RNA is independently replicated in Arlington protoplasts? The answers to these questions could give information on the inhibitory activity in Arlington.

Bruening: We have not done experiments with purified RNAs as inoculum to Arlington protoplasts and have not assayed for replicating RNA, so we don't know about RNA replication. We have tried to investigate whether the inhibitory activity affects processing in the B component RNA, but the results so far are ambiguous.

Goldbach: If the inhibitor is always present in Arlington cells, how do you explain the initial replication of the virus?

Bruening: We don't have evidence for viral replication in intact plants. However, in the protoplasts we do.

Goldbach: One might expect to observe some replication in the cells which are hit by the inoculated virus?

Bruening: Ken Eastwell and Mike Kiefer (Eastwell et al 1983) heavily inoculated intact plants with virus and tried to detect (−) RNA, which might be one of the first indications of replication. From one of the other cowpea lines, which gives rise to susceptible protoplasts, we saw production of full-length (−)RNA, but in Arlington seedlings that were heavily inoculated with CPMV we found no evidence for synthesis of intact $(-)$RNA, of either RNA-1 or RNA-2.

Goldbach: There should be some 'subliminal' (i.e. undetected) replication, in view of the interaction between CPSMV and CPMV.

Bruening: We must have some expression of the genetic information of CPMV, but it could be the translation of a protein. It wouldn't necessarily be RNA replication. We are dealing with the initially inoculated cells, and only a small amount of the CPMV and CPSMV RNAs enters those cells. A polyprotein translated from CPMV might have some effect on the few molecules of CPSMV RNA that are present.

Baulcombe: Have you tried making pseudorecombinants between cowpea mosaic virus and cowpea severe mosaic virus?

Bruening: We attempted that, but it does not work because the viruses are too different.

Dodds: Is the interaction between CPMV and CPSMV a 'site-specific' reaction? Where do you inoculate the second virus—on the same leaf as the first virus?

Bruening: Yes. Fernando Ponz and I coined the term 'concurrent protection' for the phenomenon observed with Arlington cowpeas and CPMV and CPSMV, because you must inoculate the two viruses simultaneously, on the same leaf, in order to observe the inhibition.

Dodds: What is the effect of inoculation of the second virus in distant leaves?

Bruening: No interference is observed when we inoculate distant leaves or at a later time.

Dodds: Does the second virus multiply?

Bruening; Yes. Fernando Ponz and Adib Rowhani have attempted the same interference experiment with tobacco ringspot virus (TobRV) and cowpea severe mosaic virus. They found that the inhibition of spread of CPSMV to trifoliate leaves was again a concurrent phenomenon.

Zaitlin: Arlington is the only cowpea variety or cultivar among the thousand or so that you have tested from which resistant protoplasts were obtained. Is there anything unusual about the genetic origin of Arlington?

Bruening: We have attempted to investigate that, but we have not been able to discover the origin of Arlington. The US Department of Agriculture plant introduction literature says only that it was derived from the USA.

Davies: Have all your experiments on protoplasts been done with mesophyll cell protoplasts, or have you studied protoplasts from epidermal and root tissue to see whether the gene is also expressed in other cell types?

Bruening: Given the way we prepare cells, the majority are seedling leaf mesophyll protoplasts. With regard to cell types, we were concerned that what is observed might be a 'subliminal infection' phenomenon. Therefore, Mike Kiefer examined Arlington protoplasts, which had been inoculated *in vitro*, with fluorescent antibodies raised against CPMV. He was interested to know whether a small number of protoplasts accumulated CPMV in specific cells to a high level and whether other cells had virtually no virus. This might account for the 1–10% yield of CPMV from Arlington cowpea protoplasts. However, he observed the same low level of CPMV capsid antigen throughout the cell population. This result is consistent with a cell-based immunity phenomenon that is not related to subliminal infection.

Lommel: Have you checked other types of virus that proteolytically process in Arlington cowpeas, for example potyviruses, to see whether processing is inhibited?

Bruening: We have not checked potyviruses. We examined nepoviruses, which is how we decided to work on tobacco ringspot virus.

Goldbach: Does the proteinase inhibitor work on CPSMV?

Bruening: No. Nor does it work on tobacco ringspot virus, despite the fact that Arlington cowpea is immune to TobRV.

Harrison: In the interaction between tobacco ringspot and cowpea severe mosaic is there any difference if you use satellite-containing or satellite-free tobacco ringspot virus?

Bruening: In all our experiments on interference, we have been careful to use satellite-free tobacco ringspot virus. We have worked a great deal with the satellite in our laboratory, and therefore it is important to exclude it, which we did by taking advantage of a biological source known to be free of satellite RNA. We have not tested what effect satellite might have on the interference phenomenon. Every test that we have applied has shown the inoculum that we use to be free of satellite RNA.

Beachy: What is the relative virus concentration in the competition experiment you performed with tobacco ringspot virus and cowpea severe mosaic virus?

Bruening: A really strong inoculum of tobacco ringspot virus is required to produce the protective effect. We used 10 times as much TobRV (40 µg/ml) as CPSMV (4 µg/ml) to get protection. When we reduced the concentration of tobacco ringspot virus to 1.25 times the concentration of CPSMV, the protective effect was virtually lost. In contrast, the protective effect of CPMV against cowpea severe mosaic virus remained at a 1:1 mass ratio.

White: Was the protective effect of TobRV also produced when only RNA was used as virus inoculum?

Bruening: We have not yet tested RNA with the tobacco ringspot virus system, nor have we checked to see whether Arlington cowpea protoplasts are resistant to TobRV.

Goldbach: We have additional evidence that proteolytic breakdown is not a mechanism of resistance. In *Chenopodium quinoa*, a very good host for CPMV, only the coat proteins of this virus are detectably accumulated. The non-structural proteins are rapidly degraded. Therefore, in retrospect it was a good decision to work on cowpea (*Vigna unguiculata*) rather than *Chenopodium* , because we would not have found any non-structural proteins of CPMV. However, the amount of virus produced in *Chenopodium* is as high as in the cowpea plant, which demonstrates that the proteolytic breakdown apparently occurring in this plant does not lead to decreased virus multiplication.

References

Eastwell KC, Kiefer MC, Bruening G 1983 Immunity of cowpeas to cowpea mosaic virus. In: Goldberg RB (ed) Plant molecular biology. UCLA (Univ Calif Los Ang) Symp Mol Cell Biol New Ser 12:201–211

Franssen H, Goldbach R, van Kammen A 1984 Translation of bottom component RNA of cowpea mosaic virus in reticulocyte lysate: faithful proteolytic processing of the primary translation product. Virus Res 1:39–49

Goldbach R, Krijt J 1982 Cowpea mosaic virus-encoded proteinase does not recognize primary translation products of mRNAs from other comoviruses. J Virol 43:1151–1154

Gustafson G, Ryan CA 1976 Specificity of protein turnover in tomato leaves. J Biol Chem 251:7004–7010

Ryan CA 1973 Proteolytic enzymes and their inhibitors in plants. Annu Rev Plant Physiol 24:173–196

Resistance mechanisms of tobacco mosaic virus strains in tomato and tobacco

Masamichi Nishiguchi and Fusao Motoyoshi

National Institute of Agrobiological Resources, Tsukuba Science City, Yatabe, Ibaraki 305, Japan

Abstract. Ls1 is a temperature-sensitive (ts) strain of tobacco mosaic virus (TMV) which was isolated from a culture of a wild-type strain, L. The temperature sensitivity of this mutant did not show any host specificity, when assayed in six systemic hosts and four local lesion hosts. In experiments using protoplasts and leaf discs of tomato the temperature sensitivity was shown to be associated with a defect at the non-permissive temperature in the ability of virus to move from cell to cell. Microscopic observation of Ls1-inoculated tomato leaves revealed that virus had multiplied mainly in separate single epidermal cells without spreading to neighbouring cells. This malfunction of virus movement may be due to one amino acid substitution in a TMV-encoded 30K protein induced by a mutation, as suggested by several molecular biological studies. Another mutant, $L_{11}A$, which produces small lesions on local lesion hosts, is characterized by a low level of production of 30K protein mRNA. These results indicate that the 30K protein is important for the spread of virus from cell to cell.

Resistance mechanisms against TMV in tomato, specified by *Tm-2* and *Tm-2²* genes, were investigated by fluorescent-antibody staining of virus-inoculated epidermis. In the resistant tomato lines separate single cells were mainly infected when the epidermis was inoculated with L. This was in contrast to the systemic spread of virus from primary infected cells to neighbouring cells in the epidermis of the susceptible line inoculated with L or of the Tm-2 line inoculated with Ltb1 (a mutant which breaks resistance). In these studies it is possible that interactions between gene products of the virus strain and its host are directly or indirectly responsible for expression of a type of virus resistance of plants.

1987 Plant resistance to viruses. Wiley, Chichester (Ciba Foundation Symposium 133) p 38–56

In the process of virus infection, virus in the primary infected cell moves to neighbouring cells through plasmodesmata and spreads through the conducting tissue (phloem or xylem) to the whole plant. Active cell-to-cell movement of virus has not really been established; rather, virus movement has been assumed to be a passive process, depending entirely on the passage of cytosol through plasmodesmata between two cells. However, recent studies on temperature-sensitive (ts) mutants of tobacco mosaic virus (TMV) have thrown light on this phenomenon. The accumulating evidence favours the hypothesis that virus itself actively moves from cell to cell.

In this paper we describe the behaviour of Ls1, a ts strain of tobacco mosaic virus, in tomato and tobacco, and some aspects of *Tm-2* and *Tm-2²* genes (TMV resistance genes of tomato), and we discuss host resistance mechanisms to TMV.

Behaviour of a temperature-sensitive strain of TMV (Ls1) in tomato and tobacco

Nishiguchi et al (1978) isolated a ts strain, Ls1, from a culture of TMV L, a wild tomato strain of TMV (Oshima et al 1964). Virus multiplication in leaf discs from several plant species was compared between Ls1 and L at the two different temperatures (22 and 32 °C) (Table 1). Ls1 was able to multiply very much like L at a permissive temperature (22 °C). At a restrictive temperature (32 °C) it multiplied only to low levels, whereas L multiplied to the same extent as at 22 °C. Thus Ls1 is temperature sensitive (ts) and its temperature sensitivity seems to be independent of the host plants, although it is not clear that *Cyptotania* and *Salvinia* can be infected with Ls1. Although Ls1 and L both induced symptoms in tomato and tobacco plants, those produced by Ls1 were milder than those produced by strain L. Local lesions induced by Ls1 on necrosis-responding varieties and wild species of tobacco such as *Nicotiana tabacum* cultivars Xanthi nc and Xanthi NN, *N. glutinosa* and *N. sylvestris* were smaller than those induced by L. Table 2 shows the yields of virus in protoplasts inoculated with virions or RNA. The temperature sensitivity of Ls1 was not expressed in protoplasts. These results suggest two alternative explanations for the temperature sensitivity; one possibility is that Ls1 can multiply in cells already entered but cannot move from there to the adjacent cells; the other possibility is that an isolated protoplast is physiologically different from a cell in a tissue, so that the protoplast can permit virus to multiply but the cell within a tissue cannot.

Ls1 was investigated further to determine which explanation was correct:

(1) Tomato plants were inoculated with Ls1 and then incubated at a permissive temperature for one day in order to permit virus to spread. The leaf discs from the plants were prepared and incubated at either 22 or 32 °C for a further 12 hours. Protoplasts were isolated from these discs and assayed for infection level using fluorescent-antibody staining. The number of infected protoplasts did not increase during the additional 12 hours of incubation at 32 °C but infectivity increased. However, both the number of infected protoplasts and infectivity increased at 22 °C.

TABLE 1   Effect of temperature on multiplication of Ls1 and L in leaf discs from various plant species

Expt.	Plant species	Incubation time (days)	Inoculum (mg/ml)	Ls1 strain of TMV			L strain of TMV		
				32°C	22°C	Control[a]	32°C	22°C	Control
1	Lycopersicon esculentum c.v. Fukuju no. 2	3	0.5	72	1.6×10^5	4.5×10	2.2×10^5	1.2×10^5	87
	Nicotiana tabacum c.v. Samsun	3	0.5	86	8.8×10^4	6.7×10	1.2×10^5	1.1×10^5	95
2	Petunia violacea	6	0.2	9.5	9.6×10^2	13	8.1×10^4	1.4×10^4	13
	Chenopodium murale	4	0.2	2.2	2.8×10^3	0.0	3.1×10^2	4.2×10^3	0.0
	Cryptotaenia canadensis var. japonica	6	0.5	2.3	7.3	8.9	85	1.1×10^2	2.6
	Salvia officinalis	6	0.5	0.0	6.0	8.4	36	42	5.4

Leaf discs (14 mm diameter) were prepared from leaves inoculated with virus and incubated in culture medium under continuous illumination. The homogenate of leaf discs was used as the inoculum after appropriate dilution. The values indicate the relative number of lesions on tobacco (cv. Xanthi nc), defined as:

$$\text{Total no. of lesions} \times \text{dilution factor}$$

$$\overline{\text{Total no. of lesions with standard suspension of L } (0.1 \; \mu g/ml)}$$

Experiment 1 is modified from Table 1 in Nishiguchi et al (1978).

[a] Samples taken before incubation.

TABLE 2   Effect of temperature on the infection of tomato leaf mesophyll protoplasts with Ls1 virions or Ls1 or L RNA

Temperature (°C)	Inoculum	Infection level (%)	Infectivity
32	Ls1 RNA	26	184
	Ls1 virions	71	1444
	L RNA	24	384
22	Ls1 RNA	11	57
	Ls1 virions	21	229
	L RNA	12	55

Tomato leaf mesophyll protoplasts were inoculated *in vitro* with virions or RNA in the presence of poly-L-ornithine and cultured for two days. The level of infection was determined by staining protoplasts with TMV-specific fluorescent antibody. The homogenate of protoplasts was inoculated into tobacco (cv. Xanthi nc) after appropriate dilution. The infectivity was defined as:

$$\frac{\text{Total no. of lesions} \times \text{dilution factor}}{\text{Total no. of lesions with standard suspension of L } (0.1\ \mu g/ml)}$$

(2) The virus growth curve of the leaf discs was compared to that of protoplasts. Tomato plants were inoculated with Ls1 and incubated for one day at 20–25 °C. Leaf discs were then prepared and incubated for an appropriate period (1–3 days) at either 22 or 32 °C. The virus growth curve of Ls1 at 32 °C was found to show one-step growth, similar to that of Ls1 in protoplasts, and the growth curve of Ls1 at 22 °C was the same as that of L at either temperature.

(3) Nishiguchi et al (1980) reported further observations on the behaviour of Ls1 in tomato epidermis. Leaves whose lower epidermis had been inoculated with Ls1 and incubated at either 22 or 32 °C were stained with TMV-specific fluorescent antibody. The majority of the stained cells were single separate cells in epidermis incubated at 32 °C (Fig. 1a), whereas groups of cells were stained in epidermis at 22 °C. At the latter temperature, the longer the epidermis was incubated, the larger were the groups of infected cells.

From these results it was concluded that Ls1 can multiply in primary infected cells but cannot move from cell to cell at a restrictive temperature.

Taliansky et al (1982b) showed that Ls1 could spread in tomato leaves that had been infected with either wild TMV or potato virus X (PVX), suggesting that related (TMV) or unrelated (PVX) virus could complement Ls1 in its cell-to-cell movement function. It is interesting that the ability to complement the defect of Ls1 is not specific to TMV. It may be that one virus can be complemented by another virus but not by a third one.

Leonard & Zaitlin (1982) showed that an alteration in Ls1-encoded proteins was found only in the 30K protein when coat, 30K, 130K and 180K proteins were compared with those encoded by L. The nucleotide sequence of

30K protein was compared between Ls1 and L. Only one amino acid change was found in this region (Ohno et al 1983). These results suggest that the 30K protein may be involved in virus movement.

It is well known that RNA-1 of tobacco rattle virus (TRV) can multiply and spread in a host plant (Harrison & Robinson 1986). RNA-1 encodes three proteins (29K, 140K and 170K). The 170K protein is a readthrough product of 140K. It is intriguing that the nucleotide sequence of 29K has significant local homology with that of the 30K protein of TMV (Boccara et al 1986), suggesting that the 29K protein may also function to regulate virus movement.

In brome mosaic virus (BMV), cowpea chlorotic mottle virus (CCMV) and alfalfa mosaic virus (AlMV), it is assumed that 3a protein, one of the two proteins encoded by RNA-3, is a candidate for this cell-to-cell movement function (Hull & Maule 1986, van Vloten-Doting 1986).

The 30K protein of TMV and its mRNA were found to be synthesized transiently in an early stage of infection and were not detected later (Watanabe et al 1984). This protein is suggested to be localized in the nuclei (Watanabe et al 1986). T. Godefroy-Colburn et al (unpublished paper, EMBO workshop, 6 July 1986) found, using an immunoblotting method, that P3 (3a) protein encoded by RNA-3 of AlMV was transferred from the membrane fraction to the cell wall. Shalla et al (1982) reported that the number of plasmodesmata was decreased in the cells infected with Ls1 that were incubated at a restrictive temperature. The possibility can be raised that this sort of protein functions by interacting directly with a cell wall-associated fraction including plasmodesmata, through which an infective entity might move from cell to cell.

According to Atabekov & Dorokhov (1984) there are at least three types of ts mutations which give rise to the ts phenotype associated with retarded viral transport: type 1 ts mutations affecting the synthesis of genomic RNA; type 2 ts mutations affecting the synthesis of genomic RNA coding for the putative virus-specific transport protein; and type 3 ts mutations affecting the transport protein itself. Dawson & White (1978, 1979) reported the temperature-sensitive mutants, III_2-35 and IV-35, which belong to type 1. There is no direct evidence that the 30K protein of Ls1 is temperature sensitive. However, the Ls1 mutation is thought to be of type 3. Ni2519, reported by Jockusch (1968), is also of this type, although Ni2519 has an additional ts defect of virus assembly (Taliansky et al 1982a). So far there have been no reports of temperature-sensitive mutants of type 2. Recently Watanabe et al (1987) showed that an attenuated strain, $L_{11}A$, of TMV produced smaller amounts of 30K protein and its mRNA in protoplasts than those produced by its parent wild strain, L. In contrast, other proteins (coat, 130K and 180K) were shown to be produced by $L_{11}A$ to a similar extent to those produced by L. $L_{11}A$ induced smaller local lesions than did L (Nishiguchi & Oshima 1977), suggest-

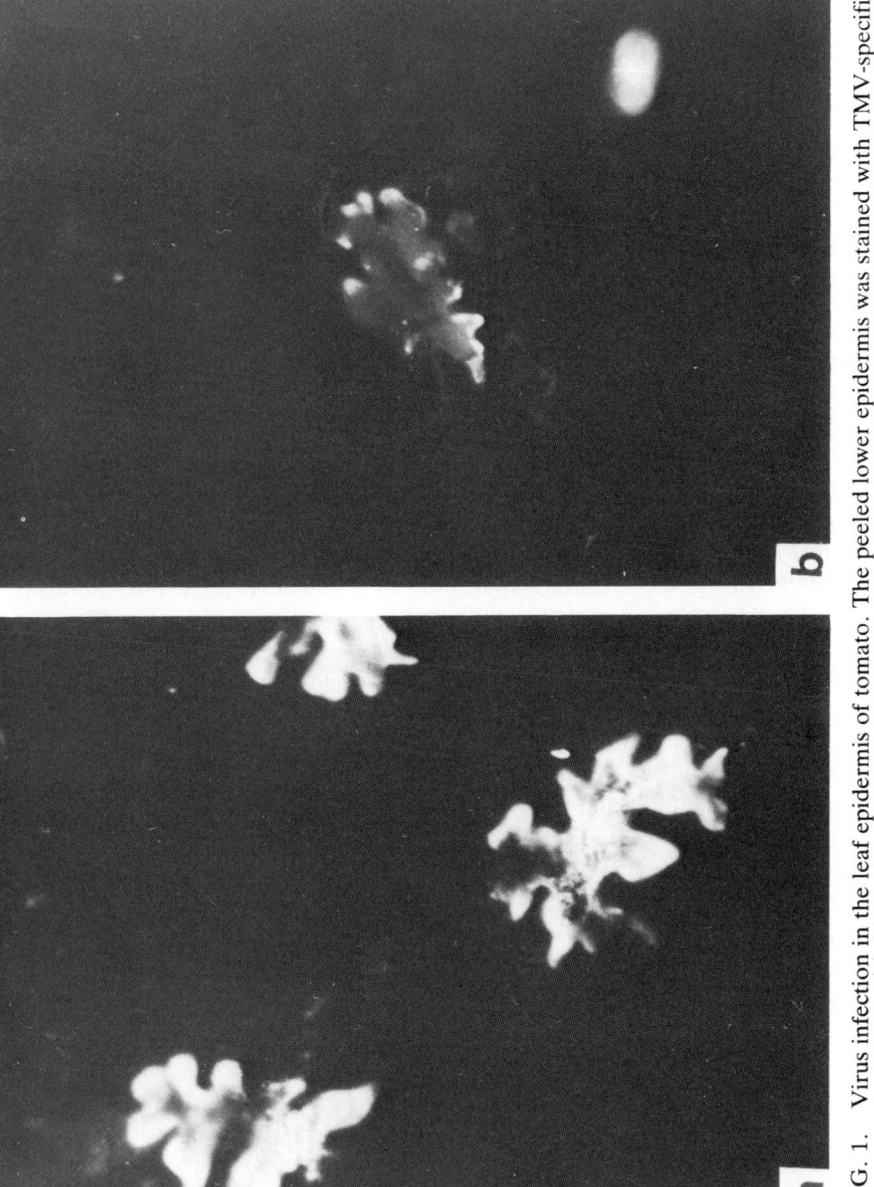

FIG. 1. Virus infection in the leaf epidermis of tomato. The peeled lower epidermis was stained with TMV-specific fluorescent antibody. (a) The epidermis of tomato of genotype +/+ (cv. Fukuju no.2) was inoculated with Ls1 and cultured at 32 °C. (b) The epidermis of tomato of genotype $Tm-2^2/Tm-2^2$ (cv. GCR 267) was inoculated with L and cultured at 25 °C.

ing that $L_{11}A$ may move slowly from cell to cell. $L_{11}A$ is not temperature sensitive, but so far as virus movement goes, it is thought to be a mutant which is partially defective in producing 30K protein mRNA. This example indicates that the 30K protein can affect virus cell-to-cell movement quantitatively, because its amino acid sequence does not differ from that of L (Nishiguchi et al 1985).

Resistance by Tm-2 and Tm-2² genes of tomato to TMV infection

There are three resistance genes to TMV infection, *Tm-1, Tm-2* and *Tm-2²*. Pelham (1972) classified TMV strains according to these resistance genes. The *Tm-1* gene expresses resistance to TMV in protoplasts but *Tm-2* and *Tm-2²* do not (Motoyoshi & Oshima 1975, 1977). The resistance mechanisms coded for by *Tm-2* and *Tm-2²* are therefore thought to operate at the level of the plant tissue.

Taliansky et al (1982c) showed that tomato plants carrying *Tm-2* permitted TMV to spread from the conducting tissue into the mesophyll only when they had been previously infected with PVX. This suggested that the *Tm-2* gene functioned to inhibit transfer of TMV from conducting tissue to mesophyll and that PVX could complement the defect of a TMV mutant in its transport function.

F. Motoyoshi (unpublished paper, 6th International Congress of Virology, 1–7 September 1984) isolated a mutant, Ltb1, from an L-inoculated tomato plant whose genotype was *Tm-2/Tm-2*. L is a group 0 and Ltb1 a group 2 strain, according to Pelham's classification. *Tm-2* tomato plants are susceptible to Ltb1 but resistant to L. Using these two virus strains and two kinds of tomato plants of genotype *Tm-2/Tm-2* and *Tm-2²/Tm-2²*, we investigated virus behaviour in the primary infection site of the leaf epidermis (Fig. 1).

In the experiment where *Tm-2* or *Tm-2²* plants were inoculated with L, single cells were observed to be stained with TMV-specific fluorescent antibody (Fig. 1b), suggesting that L could multiply in the primary infected cells. On the other hand, groups of cells were stained in Ltb1-inoculated epidermis of *Tm-2* plants. Single cells were stained in Ltb1-inoculated epidermis of *Tm-2²* plants. These results indicate that the resistance mechanisms determined by *Tm-2* and *Tm-2²* genes operated to inhibit virus movement from cell to cell, although it cannot be excluded that these genes pleiotropically suppress virus multiplication, too.

Conclusions

From our studies on a temperature-sensitive strain of TMV, Ls1, and other mutants, the phenomenon of movement of virus from cell to cell has been elucidated as a function encoded by the viral genome. It is possible that the

30K protein coded for by the TMV genome is involved in this process, both qualitatively and quantitatively, but it is not known how the 30K protein brings about the movement of virus from one cell to another. On the other hand, Tm-2 and Tm-2^2 genes are also found to regulate virus spread from the primary infected cell to neighbouring cells. There is no information on the Tm-2 and Tm-2^2 genes yet at the molecular level. It can, however, be concluded that the blocking of cell-to-cell virus movement should be considered as one resistance mechanism and that the first event in triggering the induction of resistance may be the interaction, directly or indirectly, between host- and virus-encoded products, presumably the 30K protein specified by the TMV genome and target factor(s) in the host.

Acknowledgements

The authors thank Drs N. Oshima, Y. Kiho, Y. Okada, and Y. Watanabe for collaborative work on TMV strains.

References

Atabekov JG, Dorokhov YuL 1984 Plant virus-specific transport function and resistance of plants to viruses. Adv Virus Res 29:313–364

Boccara M, Hamilton WDO, Baulcombe DC 1986 The organization and interviral homologies of genes at the 3' end of tobacco rattle virus RNA 1. EMBO (Eur Mol Biol Organ) J 5:223–229

Dawson WO, White JL 1978 Characterization of a temperature-sensitive mutant of tobacco mosaic virus deficient in synthesis of all RNA species. Virology 90:209–213

Dawson WO, White JL 1979 A temperature-sensitive mutant of tobacco mosaic virus deficient in synthesis of single-stranded RNA. Virology 93:104–110

Harrison BD, Robinson DJR 1986 Tobraviruses. In: Van Regenmortel MHV, Fraenkel-Conrat H (eds) The plant viruses: The rod-shaped plant viruses. Plenum Press, New York & London, vol 2:339–369

Hull R, Maule AJ 1986 Virus multiplication. In: Francki RIB (ed) The plant viruses: Polyhedral virions with tripartite genomes. Plenum Press, New York & London, vol 1:83–115

Jockusch H 1968 Two mutants of tobacco mosaic virus temperature-sensitive in two different functions. Virology 35:94–101

Leonard DA, Zaitlin M 1982 A temperature sensitive strain of tobacco mosaic virus defective in cell-to-cell movement generates an altered viral-coded protein. Virology 117:416–424

Motoyoshi F, Oshima N 1975 Infection with tobacco mosaic virus of leaf mesophyll protoplasts from susceptible and resistant lines of tomato. J Gen Virol 29:81–91

Motoyoshi F, Oshima N 1977 Expression of genetically controlled resistance to tobacco mosaic virus infection in isolated tomato leaf mesophyll protoplasts. J Gen Virol 34:499–506

Nishiguchi M, Oshima N 1977 Differentiation of a tomato strain of tobacco mosaic virus from its attenuated strain by the lesion type. Ann Phytopathol Soc Jpn 43:55–58

Nishiguchi M, Motoyoshi F, Oshima N 1978 Behaviour of a temperature sensitive strain of tobacco mosaic virus in tomato leaves and protoplasts. J Gen Virol 39:53–61

Nishiguchi M, Motoyoshi F, Oshima N 1980 Further investigation of a temperature sensitive strain of tobacco mosaic virus: its behaviour in tomato leaf epidermis. J Gen Virol 46:497–500

Nishiguchi M, Kikuchi S, Kiho Y, Ohno T, Meshi T, Okada Y 1985 Molecular basis of plant viral virulence: the complete nucleotide sequence of an attenuated strain of tobacco mosaic virus. Nucl Acids Res 13:5585–5590

Ohno T, Takamatsu N, Meshi T, Okada Y, Nishiguchi M, Kiho Y 1983 Single amino acid substitution in 30K protein of TMV defective in virus transport function. Virology 131:255–258

Oshima N, Goto T, Sato R 1964 A strain of tobacco mosaic virus (TMV-L) isolated from tomato. Res Bull No 83 Hokkaido Natl Agric Expt Sta, p 87–99

Pelham J 1972 Strain-genotype interaction of tobacco mosaic virus in tomato. Ann Appl Biol 71:219–228

Shalla TA, Peterson LJ, Zaitlin M 1982 Restricted movement of a temperature sensitive virus in tobacco leaves is associated with a reduction in number of plasmodesmata. Virology 60:355–358

Taliansky ME, Atabekova TI, Kaplan IB, Morozov SYu, Malyshenko SI, Atabekov JG 1982a A study of TMV ts mutant Ni2519. I. Complementation experiments. Virology 118:301–308

Taliansky ME, Malyshenko SI, Pshennikova ES, Kaplan IB, Ulanova EF, Atabekov JG 1982b Plant virus-specific transport function. I. Virus genetic control required for systemic spread. Virology 122:318–326

Taliansky ME, Malyshenko SI, Pshennikova ES, Atabekov JG 1982c Plant virus-specific transport function. II. A factor controlling virus host range. Virology 122:327–331

van Vloten-Doting L 1986 Virus genetics. In: Francki RIB (ed) The plant viruses: Polyhedral virions with tripartite genomes. Plenum Press, New York & London, vol 1:83–115

Watanabe Y, Emori Y, Ooshika I, Meshi T, Ohno T, Okada Y 1984 Synthesis of TMV-specific RNAs and proteins at the early stage of infection in protoplasts: transient expression of the 30K protein and its mRNA. Virology 133:18–24

Watanabe Y, Ooshika I, Meshi T, Okada Y 1986 Subcellular localization of the 30K protein in TMV-inoculated tobacco protoplasts. Virology 152:414–420

Watanabe Y, Morita N, Nishiguchi M, Okada Y 1987 Attenuated strains of tobacco mosaic virus: reduced synthesis of a viral protein with a cell-to-cell movement function. J Mol Biol 194:699–704

DISCUSSION

Harrison: Dr Nishiguchi, have you determined the 30K protein sequence in the Ltb1 strain of TMV?

Nishiguchi: No, but we hope to obtain the sequence soon.

van Vloten-Doting: Have you any information on the nucleotide sequence of the attenuated strain, $L_{11}A$, just preceding the initiation site for the subgenomic RNA for 30K?

Nishiguchi: We know the nucleotide sequences of both L and $L_{11}A$. Ten point mutations have been found over the whole genome. Only three of these point mutations resulted in amino acid substitutions. These were located in the region coding for the 126K protein. You are really asking about the mechanism for the production of subnormal amounts of 30K mRNA. There is no direct evidence for that. One possibility is that the change in the 126K protein may result in the reduced production of 30K protein messenger RNA.

van Vloten-Doting: Are there only changes in the 126K protein cistron, and none in the sequence preceding the subgenomic 30K mRNA transcription initiation site?

Beachy: We don't know where the initiation site is in $L_{11}A$ TMV, so we do not know which changes are important. It is possible that the changes near the end of the 183K region could be important.

Zimmern: If the $L_{11}A$ mutation does work by changing p126, because p126 is required for subgenomic mRNA synthesis, it seems strange that the synthesis of the coat protein subgenomic RNA does not also decrease with time, as for the 30K one.

van Vloten-Doting: The recognition signals may be different and even when the same protein plays a role in *trans*-activation it can have a different affinity for the two different signals after mutation.

Loebenstein: You reported recently that in this attenuated strain ($L_{11}A$) there is overproduction of the 165K protein (Kiho & Nishiguchi 1984). Have you any explanation or any suggested mechanism?

Nishiguchi: We described the subnormal production of 30K protein of $L_{11}A$, but we did not find the overproduction of 165K protein in this protoplast system. The data you mention are from experiments at the level of plant tissue, not protoplasts. At present we have no general hypothesis for these. More work is needed before we can interpret these results from different systems.

Zaitlin: We have worked on purification of the TMV replicase from tobacco (Young et al 1987). We have a solubilized replicase which contains about equimolar amounts of 183K and 126K protein. In addition we believe that there is a 54K protein, which is a readthrough in the 183K region. We have reported an mRNA for it (Sulzinski & Zaitlin 1982), and we have Western blot evidence that the 54K does exist (unpublished), in contrast to what was found by Okada's group in Japan (Saito et al 1986). It also seems to end up in a replicase fraction.

Goldbach: Dr Nishiguchi, you looked for differences in the 30K protein and found only one point mutation giving rise to one change of amino acid. You state that the temperature-sensitive (ts) effect arises from this change in protein sequence. Can you exclude the possibility that the various point mutations affect the secondary structure of the RNA, so that the ts effect is at the RNA level? Point mutations might, for instance, result in decreased stability of the RNA which is then readily broken down during transport.

Nishiguchi: That is possible, but we favour the amino acid change as the explanation, based on the comparison of the 30K sequence among different viruses and TMV strains. We have no data for an effect of silent point mutations, so we cannot exclude the possibility that silent point mutations may be responsible for virus cell-to-cell movement.

van Vloten-Doting: Have you tried to isolate revertants of Ls1? Can you retrieve something that still carries the original mutation but has a compensating mutation elsewhere in the genome?

Nishiguchi: We have some revertants, but we have never checked on that.

van Vloten-Doting: What are the phenotypes of these revertants? Are they just like the wild-type TMV?

Nishiguchi: The way we isolate revertants is to use local lesion hosts, such as Xanthi nc, and subject them to a temperature shift treatment. We check the local lesion type for revertants and find an incidence of about 0.2–0.01% of naturally occurring revertants (without mutagen treatment).

van Vloten-Doting: Have they been sequenced?

Nishiguchi: No.

Hohn: Can you say more about your studies on the spreading of virus from leaf to leaf? Is this movement also dependent on the 30K protein?

Nishiguchi: I don't think so. Dr Zaitlin may like to answer that question but, so far as I know, long-distance virus movement requires coat protein. Dr Zaitlin's group has isolated a defective TMV strain which does not produce mature virions because of its abnormal coat protein, and it is unable to spread from inoculated tobacco leaves to higher leaves. Therefore mature virions may be necessary for long-distance virus movement (Siegel et al 1962).

Zaitlin: In recent work of Okada's group, constructs were made which don't have coat protein (Takamatsu et al 1987). They behave just like the defective mutants that Dr Nishiguchi described. They move from cell to cell: they don't move long distances. This implies either that they cannot get into the vascular system or that they are labile without a coat protein and are destroyed in the vascular system.

Harrison: There are also some TMV strains, from tomato, for example, which don't move long distances but do have coat proteins.

Zaitlin: They may well have some cell-to-cell movement defect, rather than a vascular problem.

Goldbach: It may be that within the TMV RNA construct made by Okada's group (containing the CAT gene) the secondary structure is greatly altered and the stability of the RNA is diminished. Perhaps the RNA can move long distances as long as it is stable enough, as shown for some strains of TMV. As soon as larger deletions or insertions are introduced into the RNA, its stability is destroyed.

Zaitlin: The constructs in which the CAT gene was substituted also moved only from cell to cell, but not long distances.

Goldbach: That is correct, but it is not a very subtle mutation. It would be worthwhile to make a single point mutation in the coat protein gene which does not change the RNA structure and study what happens. As this mutant would no longer make coat protein, it may be established whether coat protein is indeed necessary for long-distance movement.

van Vloten-Doting: In Sarkar's mutant of TMV (which does not have coat protein: see Sarkar & Smitamana 1981) the location of the mutation is not known, but it is a naturally occurring mutant which will not involve a big deletion, and that too does not move long distances.

Zaitlin: Many of the defective mutants that we obtained involved point mutations (Zaitlin & McCaughey 1965). They result in non-functional coat proteins. Therefore the RNA was essentially unchanged. The viruses do not move long distances, but only from cell to cell.

Gianinazzi: Dr Nishiguchi, from your results one can speculate that the resistance arises from the ability of the host to stop the synthesis of the 30K protein. In fact during the hypersensitive reaction to TMV in tobacco, the virus replicates over several cycles and then multiplication is halted. Therefore the inhibition of replication may be the result of the action of the host on the synthesis of the 30K protein.

Fritig: In our laboratory, Thérèse Godefroy-Colburn and Christiane Stussi-Garaud have shown for alfalfa mosaic virus (AlMV) that the corresponding protein (P3) is incorporated into the cell wall (Godefroy-Colburn et al 1986, Stussi-Garaud et al 1987), as Dr Nishiguchi mentioned. By making point (micro) inoculations, staining the viral coat protein with antibodies and using an electron microscope, it was possible to visualize the spread of the virus, and show that this incorporation occurs at the leading edge of the infection. They now have similar biochemical data with TMV. As I discuss in my paper (Fritig et al, this volume), during the hypersensitive response to any parasite there are important changes in the cell wall. Much material is deposited there, and it might be that the 30K protein can no longer be incorporated in the cell wall. This would slow down the cell-to-cell movement of the virus and progressively lead to its localization.

Harrison: It would seem important to attempt to locate the 30K protein in cells of plants infected with Dr Nishiguchi's mutants and grown at different temperatures.

Zaitlin: Recently Dr Okada's group suggested that the 30K protein was synthesized in the nucleus (Watanabe & Okada 1986). Is that still believed?

Nishiguchi: I do not know whether they still believe that. It has been shown by T. Godefroy-Colburn et al that the 29K protein of AlMV moves from the cytosol to the cell wall-associated fraction. Therefore, because the 30K protein of TMV is thought to correspond to the 29K protein of AlMV, the target site or location of the 30K protein might be in the cell wall.

Fraser: In heterozygous tomato plants with the *Tm-2 nv* resistance gene there is sometimes a necrotic reaction to the virus. Have you done fluorescent staining to see if the virus is restricted to single cells or clumps?

Nishiguchi: Yes. In heterozygous tomato plants with the *Tm-2 nv* gene, we observed increased numbers of cell groups which were stained with the fluorescent antibody to TMV. Therefore we presume that the gene dosage affects the virus cell-to-cell movement in this case.

Fraser: Is it possible to say whether all cells were staining with a similar intensity, or was there evidence for less virus antigen away from the centre of the infected group?

Nishiguchi: In this system it is difficult to determine fluorescence quantitatively, so it is not possible to make that distinction.

Zimmern: There is a large body of work on the TMV mutant Ni2519 which is relevant to the work on Ls1 described by Dr Nishiguchi. Ni2519 was isolated some time ago in the laboratory of Drs Wittmann and Melchers at Tübingen (Jockusch 1966). It has a complex phenotype with at least three deficiencies. The selected defect is a temperature-sensitive deficiency in local lesion spreading. There is also an unselected temperature-sensitive deficiency in assembly, discovered much later (Bosch & Jockusch 1972), which is due to activation of a spurious assembly origin within the coat gene leading to abortive double initiation (Taliansky et al 1982a, Kaplan et al 1982). In addition, there is a reported deficiency in its host range—it forms local lesions instead of systemically infecting tobacco varieties such as *N.sylvestris* carrying the *N'* gene (Taliansky et al 1982b).

We have looked for mutations in Ni2519 by comparison with its parental strain, A14, using tryptic peptide mapping and nucleotide sequencing, but we have had difficulty in unscrambling the relative contributions of these three deficiencies. There is a point mutation in the 30K gene on the 5′ side of the assembly origin on Ni2519 RNA (that is, about one kilobase from the 3′ end) where it overlaps the region encoding the most highly conserved domain in the 30K protein (Zimmern & Hunter 1983). This results in both an Arg-to-Gly mutation in the protein and a possible change in the secondary structure of the RNA close to the assembly origin. This point mutation must be involved in the temperature-sensitive phenotype of Ni2519, because revertants selected for ability to grow at the restricted temperature have that mutation reverted to the wild-type. From our tryptic peptide mapping we suspect that there may be at least one other mutation in p30, while additional point mutations elsewhere in the RNA cannot be ruled out.

There are a number of reasons, such as the high mutation rate, to be cautious about relying on point mutation data in RNA viruses. The complexity of the phenotype must also be taken into account. However, the evidence linking point mutations in p30 with temperature sensitivity in local lesion spreading is quite consistent—the point mutations independently mapped in the two mutants which have been examined (Ls1 and Ni2519) are only 10 amino acids apart, and they fall within the same conserved domain of p30. Thus our evidence provides corroboration for the work of the Japanese groups of Dr Nishiguchi and Dr Okada.

The phenotypes of the mutants, as far as the local lesion spreading phenotype is concerned, are pretty much identical. It has also recently been shown that tobacco rattle virus (TRV) has a homologous gene in a similar position on its genetic map to that of this gene in tobacco mosaic virus (Boccara et al 1986, Cornelissen et al 1986). TRV is known to be able to sustain an infection in which the RNA spreads slowly from cell to cell in the absence of the coat protein. All this tends to support the suggestion made many years ago by Milt Zaitlin and Al Siegel that there are two independent transport functions in TMV (Siegel et al 1962). This hypothesis, now stated more precisely, is that one is p30 and is responsible for short-range spread: the other is the coat protein, which is responsible for longer range spread. More recent experiments have tended to confirm the association of temperature sensitivity in local lesion spreading with p30 mutations and clear up some of the inherent doubts associated with point mutant selection and complex phenotypes. Richard Turner, at the Laboratory of Molecular Biology in Cambridge, has done a deletion analysis of the assembly origin of wild-type TMV, showing that the region in which the Ni2519 point mutation lies can be deleted without affecting *in vitro* assembly (Turner and Butler 1986). I understand that there is also some further supporting evidence from Dr Okada's lab. which confirms that the Pro → Ser mutation in Ls1 is the one involved in the local lesion-spreading phenotype (Meshi et al 1987).

Matthews: What is the history of the Ni2519 strain over the past 25 years? Has it been frozen all the time or have many other mutations occurred?

Zimmern: I got it directly from Harald Jockusch when he finished working with it. Since then it has been passaged once. We have been very careful to keep the passage number low.

White: If tobacco with the *N* gene is inoculated with this mutant at 32 °C, does an infection result?

Zimmern: We haven't done that experiment.

White: How about tobacco with the *n* gene: does the virus multiply?

Zimmern: That is how we selected the revertant. It does multiply but the virus that is recovered is not the same as the virus we inoculate. That is the reversion analysis, essentially.

White: If tobacco with the *N* gene is inoculated with wild-type TMV a local lesion results. If such lesions are left for two weeks and then put at a high temperature no virus spreads from that lesion. The virus is trapped in the lesion. You observed a smaller lesion with the Ni2519: perhaps the localization reaction is much quicker with Ni2519 than with the type strain and the mature lesion forms very quickly. Putting such lesions at 32 °C would never allow the virus to spread even if it would multiply at 32 °C. It is important to know for certain that the virus doesn't spread at 32 °C, and this would be answered by inoculating plants at 32 °C.

Zimmern: Our work doesn't address virus spreading directly—the genetic assay scores the size of the visible lesion, not the underlying infected area. However, the immunofluorescence results from Dr Nishiguchi show that virus spreading is directly affected in Ls1 (Nishiguchi et al 1978, 1979).

van Vloten-Doting: Were the revertants true revertants to the original sequence, or were they revertants with the same phenotype but different genotypes?

Zimmern: The revertants were selected in such a way that both the temperature-sensitive deficiencies would have to revert to be viable. This was making a virtue out of necessity because if one expects a mutation frequency of 10^{-4}, it is advantageous to revert both deficiencies in order to recover progeny which are stably reverted. The revertants have the wild-type sequence; they are not second site revertants.

van Vloten-Doting: That is quite surprising.

Zaitlin: In Jockusch's work (1968) there was the possibility that there were wild-type contaminants in the inoculum which eventually replicated. He didn't rule that out.

Zimmern: The sequencing was deliberately done on the RNA directly, so as to look at the population consensus sequence. Any other sequence present at the level of a few per cent or lower would be below the level of detection. However, looking at the population as a whole yields a clean answer—the mutant has a Gly residue where the wild-type and the revertant have an Arg.

van Vloten-Doting: Dr Zimmern, from this you cannot know that your revertant is a true revertant, because if there was a very low level of the original sequence present and you selected this out, you would recover that original sequence.

Zimmern: That is not the point. The question is whether that particular base change is related to the phenotype change. In fact there is another base change in the parental strain A14 relative to wild-type TMV, only three bases away from the point mutation in Ni2519, which acts as an internal control—that position does not alter through either forward or back selection of Ni2519 or its revertant.

van Vloten-Doting: That does not answer my question, which is whether your revertants are true revertants. So far as I can deduce, you cannot know this.

Zimmern: The experimental design is imposed by the system, which compels you to look at the population consensus. In any systemic infection with viruses of this kind, low level sequence variation is bound to occur because of the numbers of molecules involved. In a population of more than 10^4 molecules of virus with a genomic length approaching 10 kb and a mutation frequency of

10^{-4}, statistically each molecule is bound to differ from the population average. Since you have to use at least that many molecules to initiate an infection, it is impossible to tell whether the variant which grows out was present at a low level in the original inoculum or whether it arose during that particular passage.

van Vloten-Doting: I agree. The phenotype has reverted to the wild-type. However, is it of the same genotype or has another mutation suppressed the phenotype?

Zimmern: Operationally the point mutation that is important is the one resulting in the Gly → Arg mutation. It changes when the conditions are altered. Another way of testing the consistency of the data has been used with Ls1. When site-directed mutagenesis is performed to change the relevant base deliberately, the original phenotype is recovered (Meshi et al 1987).

van Vloten-Doting: I don't deny that if that particular base is changed another phenotype may be obtained. What interests me is whether, if the original phenotype is restored, it is the original sequence, or whether the virus can have another sequence giving the same phenotype. Is there a second site mutation?

Zimmern: We are confident there is not a second mutation in the revertant p30 because we tryptic peptide mapped the mutant and the revertant. However, tryptic peptide mapping does not give sufficient resolution in p126 or p183 to determine whether there is a second site mutation in these. We are not in a position to say whether a second site mutation could suppress the original one.

Beachy: I would like to add some more information about the Ls1 mutant discussed by Dr Nishiguchi. We isolated a cDNA clone that contained an open reading frame encoding the 30K gene from the U1 strain of TMV. We placed this open reading frame in an intermediate plasmid behind the 35S promoter of cauliflower mosaic virus, and performed transformations mediated by a disarmed Ti plasmid in *Agrobacterium tumefaciens*. From the resulting transgenic tobacco plants, we obtained three sizes of RNA molecules; 1.4 kb, 1.3 kb and 1.1 kb. All the RNAs were polyadenylated and were found on large polysomes.

Do these transgenic plants make the 30K protein and does it accumulate? We made two extracts from transgenic leaves—a soluble fraction and a cell wall fraction—and examined each for the 30K protein. We also examined control plants that had been inoculated with 200 μg/ml of the U1 strain of TMV, taking the leaves three days after infection. We did a Western blot analysis with an antibody raised against a synthetic oligopeptide that represents an internal portion of the protein. The 30K protein was found in the TMV-infected control plant, but there was at least 10 times more (on a leaf fresh weight basis) of the protein in the transgenic plant than in the U1-infected control. The 30K protein was not detected in a transgenic plant that was transformed to carry only the intermediate plasmid, nor in transgenic plants expressing the TMV coat protein gene. The strongest immunoreaction was observed with the 30 kDa protein found in the soluble fraction of extracts from transformant 277 (expressing the

30 kDa gene). A soluble host protein was extracted from non-infected Xanthi tobacco plants which also reacted with the antibody. The significance of this cross-reactive material is not known.

Thus we have transgenic tobacco plants in which the TMV 30K gene is expressed. These plants were inoculated with the Ls1 strain of TMV; the plants were held for one day at the permissive temperature (22 °C), then shifted to 32 °C. Disease did not develop in control plants that lack the 30K gene but have only the intermediate plasmid. In the transgenic plant with the 30kDa gene, full disease development was seen with both the Ls1 and L strain of TMV. Unlike those of the control plant (not expressing the 30kDa gene), the lower leaves of the plants expressing the 30kDa gene were infected and systemic spread occurred.

The leaves of ten different transgenic seedlings of transformant 277 were inoculated with Ls1 and left for 10–12 days. We then monitored accumulation of virus in the inoculated leaves as well as in the upper leaves. In control plants we detected a low amount of antigenic reaction in the inoculated leaf, and no systemic movement of the virus at 32 °C. In the 277 transformants, more virus accumulated in the inoculated leaf than in the controls and systemic spread occurred. Thus, when the U1 TMV 30kDa gene is expressed in tobacco plants, the Ls1 mutant is complemented, and the results help to support the idea that the 30K gene is involved in cell-to-cell movement as well as being required for systemic spread. We have termed the 30K protein the 'Movement Protein' (MP) (Deom et al 1987).

Harrison: That is an important result. Are you doing immuno-gold studies to locate the 30K protein in cells of the transgenic plants?

Beachy: This is in progress, but we have no results yet.

van Vloten-Doting: Did you infect the transgenic plants with other viruses?

Beachy: We have tried to infect these plants with other viruses. The results are encouraging but not yet conclusive.

Zaitlin: Dr Beachy, the fact that you obtained so much 30K in these plants, relative to the case of natural infection, must say something about the control of the 30K gene. Have you any thoughts about that?

Beachy: We favour the idea that, with the encapsidation site on the subgenomic RNA that encodes the 30kDa protein, the amount of this protein produced during the infection cycle is controlled by encapsidation of the RNA.

Zaitlin: Have you made transformed plants with both the coat protein and the 30K protein?

Beachy: We have already genetically crossed plants expressing the 30kDa or the coat protein genes, and shall soon have plants expressing both gene products. We did this for two reasons: to see whether the coat protein in the transgenic plants is capable of encapsidating the mRNA encoding the 30kDa gene; and to discover which protein is more effective. When the 30K plants are infected with the L or Ls1 strain there is no cross-protection. Rather, the plants are somewhat more susceptible, apparently because the virus can move more rapidly. Therefore experiments with progeny of the genetic cross will be informative.

Nishiguchi: How about the dosage effect of 30K protein on Ls1 spreading in the transgenic plants? Have you made a comparison of transgenic plants with low and high levels of the 30K protein?

Beachy: We looked at six different transgenic plants. Initially we worked with one with a high level of expression—the 277 transformant. Another transformant (number 274) makes lower amounts of RNA and protein but still permits virus transport. Others make even lower levels of protein but we have not yet studied those, so I don't know what the required dosage for transport is.

Nishiguchi: Does symptom expression by Ls1 differ in these plants?

Beachy: Ls1 doesn't move as rapidly as L in these transgenic plants. We did the transformations with the systemic host, Xanthi, and also with Xanthi nc, so we have the size of the local lesion to monitor too. We don't have results yet on Xanthi nc plants but we should soon discover whether we observe the halo phenotype, for example, or whether the virus grows beyond the halo.

Harrison: Those results will carry us further forward. We have inoculated the RNA-1 of a nepovirus to plants infected with tobacco rattle virus. The nepovirus RNA-1 can infect protoplasts but has no obvious effect on intact plants. Tobacco rattle virus seemed unable to assist it to spread within the plant.

Beachy: Has anyone else complemented one virus transport function by another? Has anything been achieved since the review by Atabekov & Dorokhov (1984)?

Dodds: Apparently, bean golden mosaic virus (a geminivirus normally limited to the phloem tissue in bean plants) invades other tissues when unrelated viruses, such as TMV, are also present (Carr & Kim 1983).

References

Atabekov JG, Dorokhov YuL 1984 Plant virus-specific transport function and resistance of plants to viruses. Adv Virus Res 29:313–364

Boccara M, Hamilton WDO, Baulcombe DC 1986 The organisation and interviral homologies of genes at the 3' end of Tobacco Rattle Virus RNA1. EMBO (Eur Mol Biol Organ) J 5:223–229

Bosch FX, Jockusch H 1972 Temperature-sensitive mutants of TMV: behaviour of a non-coat protein mutant in isolated tobacco cells. Mol Gen Genet 116:95–98

Carr RJ, Kim KS 1983 Evidence that bean golden mosaic virus invades non-phloem tissue in double infection with tobacco mosaic virus. J Gen Virol 64:2489–2492

Cornelissen BJC, Linghorst HJM, Brederode FTh, Bol JF 1986 Analysis of the genome structure of Tobacco Rattle Virus strain PSG. Nucl Acids Res 14:2157–2169

Deom CM, Oliver M, Beachy RN 1987 The 30-kilodalton gene product of tobacco mosaic virus potentiates virus movement. Science (Wash DC) 237:389–394

Fritig B, Kauffmann S, Dumas B, Geoffroy P, Kopp M, Legrand M 1987 Mechanism of the hypersensitivity reaction of plants. In: Plant resistance to viruses. Wiley, Chichester (Ciba Found Symp 133) p 92–108

Godefroy-Colburn T, Gagey MJ, Berna A, Stussi-Garaud C 1986 A non structural protein of alfalfa mosaic virus in the walls of infected tobacco cells. J Gen Virol 67:2233–2239

Jockusch H 1966 Temperature-sensitive Mutanten des Tabakmosaikvirus. I. In-vivo verhalten. Z Vererbungsl 98:320–343

Jockusch H 1968 Two mutants of tobacco mosaic virus temperature sensitive in two different functions. Virology 35:94–101

Kaplan IB, Kozlov YuV, Pshennikova ES, Taliansky ME, Atabekov JG 1982 A study of TMV ts mutant Ni2519. III. Location of the Reconstitution Initiation Sites on Ni2519 RNA. Virology 118:317–323

Kiho Y, Nishiguchi M 1984 Unique nature of an attenuated strain of tobacco mosaic virus: autoregulation. Microbiol Immunol 28:589–599

Meshi T et al 1987 Function of the 30 kd protein of tobacco mosaic virus: involvement in cell-to-cell movement and dispensability for replication. EMBO J 6:2557–2563

Nishiguchi M, Motoyoshi F, Oshima N 1978 Behaviour of a temperature sensitive strain of tobacco mosaic virus in tomato leaves and protoplasts. J Gen Virol 39:53–61

Nishiguchi M, Motoyoshi F, Oshima N 1979 Further investigation of a temperature-sensitive strain of tobacco mosaic virus: its behaviour in tomato leaf epidermis. J Gen Virol 46:497–500

Saito T, Watanabe Y, Meshi T, Okada Y 1986 Preparation of antibodies that react with the large non-structural proteins of tobacco mosaic virus by using Escherichia coli expressed fragments. Mol Gen Genet 205:82–89

Sarkar S, Smitamana P 1981 A proteinless mutant of tobacco mosaic virus: evidence against the role of a viral coat protein for interference. Mol Gen Genet 184:158–159

Siegel A, Zaitlin M, Sehgal OP 1962 The isolation of defective tobacco mosaic virus strains. Proc Natl Acad Sci USA 48:1845–1851

Stussi-Garaud C, Garaud JC, Berna A, Godefroy-Colburn T 1987 In situ location of an alfalfa mosaic virus non structural protein in plant cell walls: correlation with virus transport. J Gen Virol 68:1779–1784

Sulzinski MA, Zaitlin M 1982 Tobacco mosaic virus replication in resistant and suscepti-ble plants: in some resistant species virus is confined to a small number of initially infected cells. Virology 121:12–19

Takamatsu T, Ishikawa M, Meshi T, Okada Y 1987 Expression of bacterial chloram-phenicol acetyltransferase gene in tobacco plants mediated by TMV RNA. EMBO (Eur Mol Biol Organ) J 6:307–311

Taliansky ME, Kaplan IB, Jarvekulg LV, Atabekova TI, Agranovsky AA, Atabekov JG 1982a A study of TMV ts mutant Ni2519. II. Temperature-sensitive behaviour of Ni2519 upon reassembly. Virology 118:309–316

Taliansky ME, Atabekova TI, Kaplan IB, Morozov SYu, Malyshenko SI, Atabekov JG 1982b A study of TMV ts mutant Ni 2519. I. Complementation experiments. Virology 118:301–308

Turner DR, Butler PJG 1986 Essential features of the assembly origin of tobacco mosaic virus RNA as studied by directed mutagenesis. Nucl Acids Res 14:9229–9242

Watanabe Y, Okada Y 1986 In vitro viral RNA synthesis by a subcellular fraction of TMV-infected protoplasts. Virology 149:64–73

Young ND, Forney JA, Zaitlin M 1987 Tobacco mosaic virus replicase and replicative structures. In: Proc John Innes Symp, September 1986. In press

Zaitlin M, McCaughey WF 1965 Amino acid composition of a non functional tobacco mosaic virus protein. Virology 26:500–503

Zimmern D, Hunter T 1983 Point mutation in the 30K open reading frame of TMV implicated in temperature-sensitive assembly and local lesion spreading of mutant Ni2519. EMBO (Eur Mol Biol Organ) J 2:1893–1900

The role of pathogenesis-related proteins

John F. Antoniw and Raymond F. White

Plant Pathology Department, Rothamsted Experimental Station, Harpenden, Herts AL5 2JQ, UK

Abstract. Pathogenesis-related (PR) proteins are plant proteins produced in response to infection by pathogens. In tobacco the PR-1 group of proteins are closely associated with virus localization and are found in greatest amount where virus spread is halted. PR proteins are also found in infected plants showing induced resistance to a second infection and are produced in response to a wide range of localized pathogens including bacteria, fungi and viruses. Several chemicals which induce resistance to virus infection and spread also induce the production of PR proteins. PR proteins serologically related to the PR-1 group of proteins in tobacco have been found in a wide range of plants, both mono-cotyledonous and dicotyledonous. The retention of common antigenic determinants in both subclasses of angiosperms suggests that PR proteins have an important, if as yet undefined, role in the response of plants to pathogens.

1987 Plant resistance to viruses. Wiley, Chichester (Ciba Foundation Symposium 133) p 57–71

Pathogenesis-related (PR) proteins were discovered independently by Giani-nazzi et al (1970) and van Loon & van Kammen (1970) in tobacco plants (*Nicotiana tabacum*) reacting hypersensitively to infection with tobacco mosaic virus (TMV). These proteins, which are coded for by the host plant genome, are not normally expressed in healthy developing plants but are produced in large amounts in response to localized infection by a wide range of pathogens, including viruses, bacteria and fungi. PR proteins have now been found in many plant species, including other *Nicotiana* species, tomato, potato, bean, cowpea, cucumber, celery, citron, *Gomphrena globosa* and *Gynura aurantiaca* (for reviews see Redolfi 1983; van Loon 1985), and therefore may be an important general response of plants to infection by pathogens. This paper is mainly concerned with PR proteins in tobacco plants, because these have been studied most intensively (for review see Antoniw & White 1983).

Not all tobaccos respond in the same way to infection by TMV. In cultivar Xanthi, TMV produces a systemic infection and multiplies and spreads from the infection sites throughout the whole plant. In cultivar Xanthi nc, which

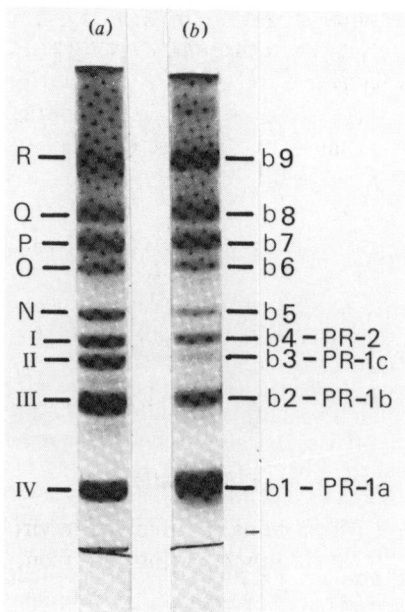

FIG. 1. Electrophoresis of extracts of TMV-infected leaves of (a) Samsun NN and (b) Xanthi nc in 10% polyacrylamide gels under non-denaturing conditions, showing nine tobacco PR proteins and the three main systems of nomenclature. (From Antoniw et al 1980 by permission of the Society for General Microbiology.)

contains the *N* gene derived from *N. glutinosa*, TMV multiplies only to a limited extent and is restricted to the areas of leaf immediately around the infection sites, which become necrotic within a few days. This is called a localized infection or hypersensitive reaction, and forms a very effective means of resistance to TMV infection.

The PR proteins were discovered using non-denaturing polyacrylamide gel electrophoresis, and comparing leaf extracts of healthy tobaccos with tobaccos showing localized or systemic infections of TMV. Several new proteins were identified in those tobaccos which localized TMV, and these were called 'b-proteins' in tobacco cultivar Xanthi nc and 'new components' in cultivar Samsun NN (Fig. 1). More recently, the name 'pathogenesis-related (PR) proteins' was suggested because the proteins were found in pathogen-infected leaves, and a unified system of nomenclature was proposed to emphasize similarities between certain proteins (Antoniw et al 1980).

PR proteins b1, b2 and b3 from Xanthi nc and components IV, III and II from Samsun NN tobacco were the first to be purified and were shown to have similar molecular masses (about 14 000 Da) but different charges. They also have similar amino acid compositions, are serologically related to each other, and so form a family of related proteins called the PR-1 group (Antoniw et al 1980). Several other PR proteins have since been purified, including b0, b1 and b3 from *N. sylvestris*, b2 from *N. tomentosiformis* and b1'' from *N. glutinosa* and *N. debneyi* (Ahl et al 1985). These have also been classed as PR-1 group proteins because of their similar molecular masses and their serological relationship to Xanthi nc PR-1a (b1). An antiserum raised against purified b4 (PR-2) from Xanthi nc has been used to show that b4, b5 and b6 form a second group of serologically related proteins which differ from the PR-1 group (Fortin et al 1985). Furthermore, PR proteins 2 (b4), N (b5), O (b6), P and Q have since been purified from Samsun NN (Jamet & Fritig 1986) and can be classed into two separate groups on the basis of size. It will be interesting to see whether, when their serological relationships have been determined, PR-P and -Q will form a third group of related PR proteins.

Are PR proteins involved in virus localization?

To investigate whether PR proteins are involved in virus localization we used an antiserum to the PR-1a (b1) protein (purified from Xanthi nc tobacco) in an $F(ab')_2$ enzyme-linked immunosorbent assay (ELISA) to follow the production of PR proteins during the localization of TMV. This very sensitive assay can detect as little as 20 pg PR-1a (Antoniw et al 1985), but the antiserum also reacts with PR-1b (b2) and PR-1c (b3) in double-diffusion tests, Western blots and in ELISA, although in the latter it appears to be more sensitive for PR-1a than PR-1b or PR-1c. The sensitivity and specificity of the assay should be borne in mind when interpreting these results. However, for the sake of simplicity, we shall refer to the measurement of PR-1 protein with this assay. Also, these results refer only to the PR-1 group of proteins. Other PR proteins may behave in different ways during TMV localization, and they still have to be investigated.

ELISA was first used to determine the levels of PR-1 protein in healthy and TMV-infected tobacco. In apparently healthy eight-week-old Xanthi nc tobacco plants we found concentrations of 0.4 (SE ± 0.4) ng PR-1/g leaf. When leaves were inoculated with water and assayed six days later the concentration was 0.88 (SE ± 1.28) ng PR-1/g leaf, showing that physical damage of the leaf during the inoculation produced no significant increase. However, similar leaves six days after inoculation with TMV, and showing a localized infection, contained 20.0 (SE ± 7.6) µg PR-1/g leaf — a dramatic 50 000-fold increase in PR-1.

The responses of tobacco to systemic and localized TMV infections were investigated using ELISA to measure the amounts of TMV and PR-1 protein in leaves during the first few days of TMV infection of Xanthi and Xanthi nc (Antoniw et al 1985) (Fig. 2). In the systemic infection of Xanthi with TMV, the virus concentration increased rapidly from the fourth day after inoculation, reaching 1 mg TMV/g leaf by the sixth day. However, the concentration of PR-1 protein did not increase significantly. In the localized infection of Xanthi nc the necrotic lesions appeared between the second and third days after inoculation and the main increase in virus was later, but reached only about 4% of the virus found in the systemic infection by the sixth day after inoculation. The observation that the main increase in virus occurred after the appearance of lesions is important because it shows that lesion formation is not the only factor involved in virus localization. The work of Weststeijn (1981) showed that restriction of virus spread occurred in living cells surrounding the lesion and that eventually the virus was trapped in the mature lesion. In contrast to what occurred during the systemic infection, the concentration of PR-1 increased dramatically from the third day after inoculation, just after the appearance of lesions. This indicates that PR-1 proteins are not involved in the initial collapse of the lesion but are produced in large amounts at the time when the increase in virus is being restricted. This suggests that PR-1 proteins could be having an influence at this time.

The ELISA was sensitive enough to detect PR-1 and TMV in different sections of a single lesion so it was used to measure the distribution of PR-1 and TMV around single local lesions, and to investigate how these distributions changed during the development and maturation of local lesions (Antoniw & White 1986). Fig. 3a shows that TMV was always most concentrated in the centre of the lesion but this concentration decreased rapidly with distance from the centre, so that none could be detected beyond 5 mm. The PR-1 protein (Fig. 3b) also first appeared in the centre of the lesion, but subsequently was much more widely distributed than TMV. By the seventh day after inoculation the distribution of PR-1 was markedly different from that of TMV. The largest amount of PR-1 was in a ring just outside the central area which contained the highest concentrations of virus. The small amount of PR-1 protein in the centre of the lesion is remarkable, but is unlikely to be an artefact since the PR-1 and TMV concentrations were measured in the same extracts, and TMV extraction was apparently unaffected by lesion development. It is possible that the cells in the central region of the lesion become necrotic before they are able to synthesize PR proteins. This distribution of the PR-1 protein with the largest concentrations at the periphery of the lesion is further evidence that they are associated with the restriction of TMV spread. It could be argued that necrotic lesion formation induces PR proteins. However, necrosis induced chemically, e.g. by HCl and H_3PO_4, does not induce PR proteins (van Loon 1977), and necrosis is not necessary for PR protein induction, because certain chemicals (e.g. aspirin) induce PR proteins without necrosis and the interspecific hybrid (*N. glutinosa* × *N. debneyi*) constitutively produces a PR-1 protein (see below).

Recently, Dumas & Gianinazzi (1986) have shown that at 15 °C TMV is localized in necrotic lesions in *N. rustica* but at 20 °C, although local necrotic

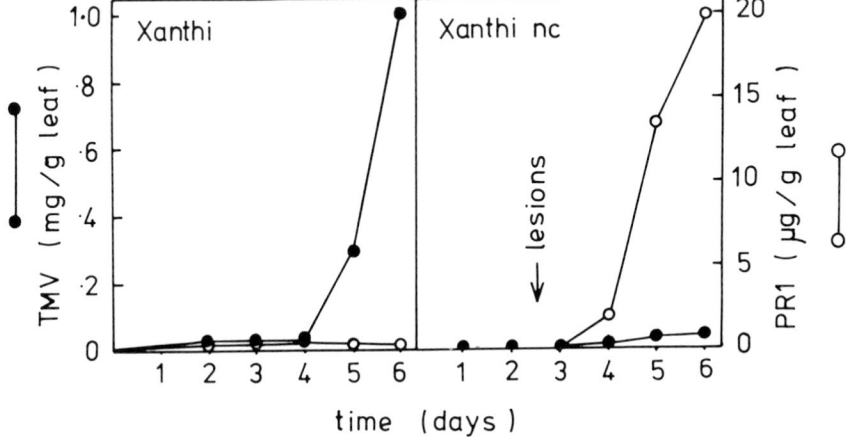

FIG. 2.   The concentrations of PR-1 protein (o—o) and TMV (•—•) determined by ELISA at various times after inoculation with TMV in systemic (Xanthi) and localized (Xanthi nc) infections. (From Antoniw et al 1985 by permission of Martinus Nijhoff Publishers.)

lesions form on the inoculated leaves, secondary necroses develop on the upper, growing leaves, showing that the virus has escaped from the primary lesions. Furthermore, PR proteins are produced at both temperatures, suggesting that they are not involved in virus localization. Nevertheless the host plant *has* recognized the virus as a pathogen and has activated its resistance mechanism(s) at 20 °C, as shown by the formation of local necrotic lesions which stop expanding 12 days after inoculation. Although TMV was detected in the midrib of the inoculated leaf, none could be found in the green areas of leaf between the necrotic lesions. In this case, although both lesions and PR proteins appeared slightly earlier at 20 °C than at 15 °C, the localization mechanism was slower at the higher temperature, because lesions stopped expanding two days later at 20 °C than at 15 °C. It is therefore possible that the virus escaped into the vascular system before localization was effective.

In conclusion, PR-1 proteins are closely associated with TMV localization in tobacco — they are produced in the right circumstances (i.e. in localized infections but not in systemic infections of TMV), at the right time, and in the right place, but are also produced under some circumstances when localization is not fully effective.

Are PR proteins involved in induced resistance?

Another aspect of the hypersensitive reaction is that other uninfected leaves on the same plant become resistant to further infection (Ross 1966). This local and systemic acquired resistance is also effective against other unrelated

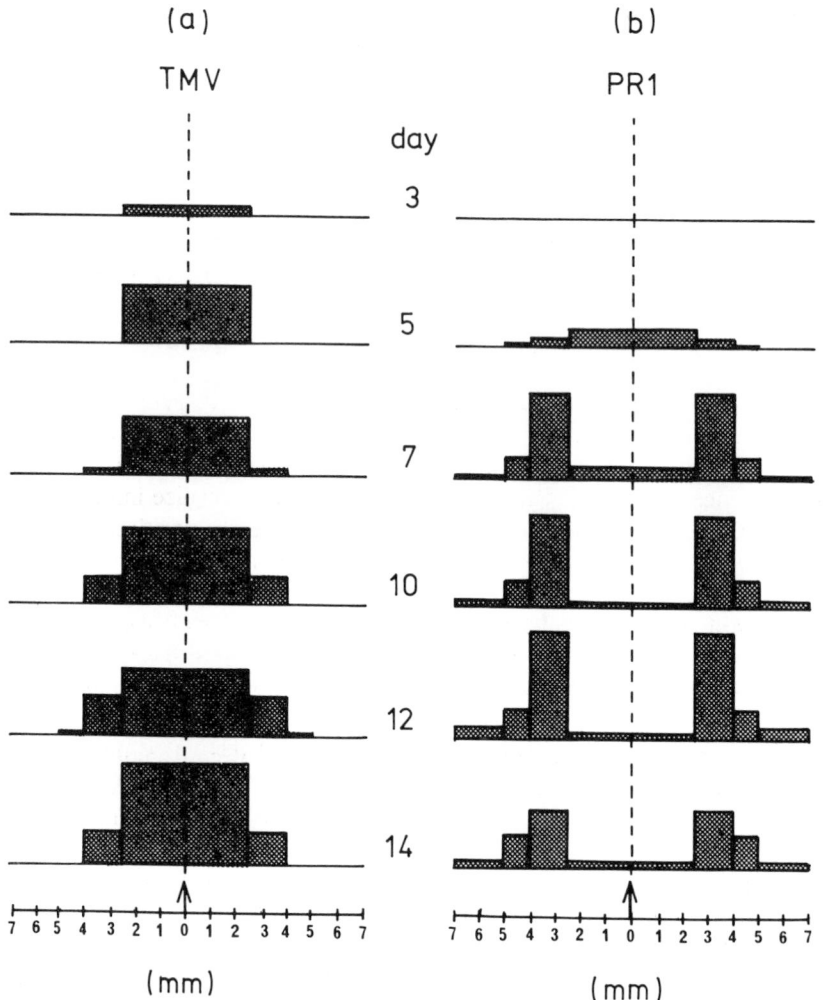

FIG. 3. TMV and PR-1 protein content of local necrotic lesions at different times after inoculation of Xanthi nc leaves with TMV. The TMV (a) and PR-1 protein (b) levels of different parts of the lesion are shown as a series of histogram plots arranged as a cross-section through the lesion. The centre of the lesion is shown as a dotted vertical line. (From Antoniw & White 1986 by permission of Martinus Nijhoff Publishers.)

pathogens. For example, in tobacco plants TMV induces systemic resistance to itself, to unrelated viruses (e.g. tobacco necrosis virus (TNV)), to fungi (e.g. *Phytophthora parasitica*) and to bacteria (e.g. *Pseudomonas tabaci*). Conversely, other pathogens which produce local necrotic lesions in tobacco, such as unrelated viruses (e.g. TNV), fungi (e.g. *Thielaviopsis basicola*) and bacteria (e.g. *Pseudomonas syringae*) induce resistance to TMV (for review see Gianinazzi 1983). A common component of tobacco's response to these different types of localized pathogens, whether viral, bacterial or fungal, is the production of PR proteins. Furthermore, these proteins also appear, in smaller amounts, in uninfected leaves showing systemic acquired resistance, which suggests that they may also be part of this response. It seems unlikely, although not impossible, that a single resistance mechanism could protect plants against such diverse pathogens. This suggests that there may be a coordinated response by the plant and that infection by one localized pathogen induces several, separate resistance mechanisms for the different types of pathogen, and different PR proteins may be involved in each of these.

The induction of systemic acquired resistance by pathogens is primarily characterized by a reduction in lesion size (Ross 1966), possibly as a result of the hypersensitive reaction occurring earlier and faster than in normal leaves. If large concentrations of PR proteins are required to localize infection, then induction of small amounts in leaves showing systemic acquired resistance would be evidence that the mechanism(s) of virus localization have been primed, so that viruses would be localized in a shorter time, resulting in smaller lesions. This model for induced resistance predicts that chemicals which induce the natural resistance or localization mechanism should also induce PR proteins. Several chemicals, including ethephon, salicylic acid, aspirin, benzoic acid, polyanions such as polyacrylic acid (see Gianinazzi 1983) and the antiviral chemicals 2-thiouracil and dioxohexahydrotriazine (White et al 1986), do induce resistance and PR proteins in the treated leaf. The induction of PR proteins in tobacco by chemicals is not due to non-specific stress, since out of a range of salts of ten metals only those of barium and manganese induced PR proteins. Many of the other salts tested were phytotoxic at high concentrations and produced chlorotic or necrotic symptoms but did not induce PR proteins (White et al 1986).

The association between the induction of resistance to TMV, measured as a decrease in lesion size, and PR proteins as a result of treatment with aspirin and polyacrylic acid has been most widely studied. TMV is localized in tobacco cultivars Xanthi nc and Samsun NN, but polyacrylic acid induces both PR proteins and resistance in Xanthi nc but neither in Samsun NN. On the other hand, aspirin induces both PR proteins and resistance in both cultivars (White et al 1983). The relationship between the concentrations of PR proteins and resistance to TMV induced by aspirin is dose-dependent (White et al 1986). Over the range 9–75 µg aspirin/ml injected into Xanthi nc leaves the levels of PR-1 protein increased and the size of the lesions produced by TMV decreased (Fig. 4).

Further evidence which supports the association of PR proteins and resistance is the discovery of an interspecific hybrid of *N. glutinosa* × *N. debneyi*

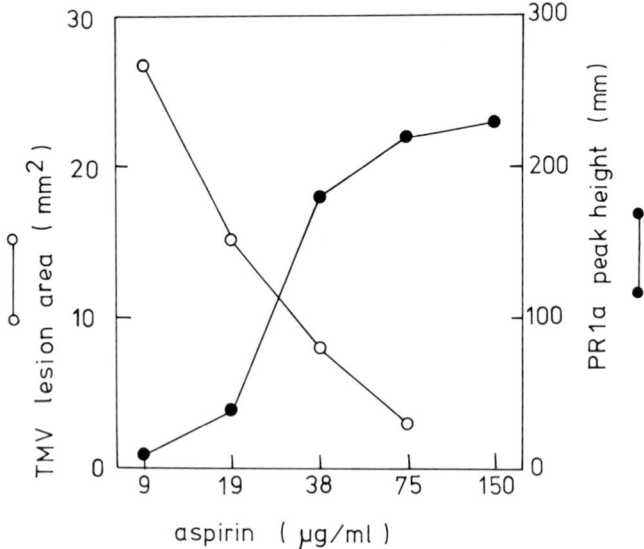

FIG. 4. The effect of injection of Xanthi nc leaves with various concentrations of aspirin on the amounts of PR-1a protein accumulated within seven days and the size of lesions formed by TMV inoculated seven days after aspirin treatment and measured another seven days later. PR-1a protein was measured as peak height (●—●) on densitometric scans of non-denaturing 10% polyacrylamide gels and lesion size was measured as lesion area (o—o). (From White et al 1986 by permission of Elsevier Science Publishers.)

which constitutively produces the b1″ PR protein and is highly resistant to infection with TMV and TNV (Ahl & Gianinazzi 1982). Furthermore, the lesions formed on the hybrid by TMV and TNV are about 93% and 56% smaller, respectively, than the lesions produced by these viruses on one of the parent plants (*N. glutinosa*), suggesting that the induced resistance is more effective with TMV than TNV.

However, not all the evidence supports the hypothesis that PR proteins are involved in systemic acquired resistance. PR-1 proteins accumulate in large amounts in apparently healthy, older, flowering tobacco plants which have senescing lower leaves. This accumulation was decreased if either the inflorescence or the senescing lower leaves were removed and, although the sizes of TMV lesions increased significantly when the inflorescence was removed, they did not change significantly when the senescing lower leaves were removed (Fraser 1981).

Fraser (1982) also related the levels of PR-1a protein present at different times after TMV infection with induced resistance and found a poor correlation. However, resistance was assessed by challenge-inoculating the leaves with TMV and measuring the size of the lesions seven days later. But, if induced resistance affects the multiplication, spread and localization of the virus, then not only is the concentration of PR proteins at the time of challenge-inoculation important; so is their concentration during the growth of the lesion before its size is estimated. This qualification is most relevant when the levels of PR proteins are already rapidly changing, during the early stages of infection.

In conclusion, most of the evidence shows that PR proteins are closely associated with, but not necessarily responsible for, induced resistance.

How widespread are the PR proteins?

Several solanaceous species produce PR proteins which are serologically related to the PR-1 group proteins of tobacco — for example, the p14 protein of viroid-infected tomato (Nassuth & Sänger 1986), and salicylic acid-treated potato and potato virus Y (PVY)-infected *Solanum demissum* (White 1983). However, with E. Rybicki and co-workers, we have shown, by using affinity-purified (monospecific) antibody in immunoelectroblots, that serologically related PR-1-type proteins are produced in a wide range of plants, including maize, barley, *Gomphrena globosa* and *Chenopodium amaranticolor*, as well as tobacco, tomato and potato (White et al 1987). The PR-1-type proteins detected were all soluble at pH 3 and had molecular masses in the range 14 000–18 000 Da. Furthermore, in all except *G. globosa* the PR-1-type protein was detected only in the pathogen-infected or salicylic acid-treated plant, and not in the untreated plant. The existence in both monocotyledonous and dicotyledonous plants of PR-1-type proteins which, in general, are expressed during infection by pathogens, and the retention of common antigenic determinants during evolution in this wide range of plants, suggests that PR proteins have an important, if as yet undefined, role in the response of plants to pathogens.

The involvement of PR proteins in resistance

Several pieces of evidence (see Antoniw & White 1983) suggest that tobacco PR proteins accumulate in the extracellular leaf space. Parent & Asselin (1984) showed that PR proteins are selectively found in the intercellular fluid of TMV-infected tobacco leaves. The PR proteins are remarkably stable, and appear to be well adapted to the extracellular environment, so their role may be related to this site of accumulation. Perhaps PR proteins outside the cell are able to affect the spread of viruses, but it is hard to see how such a mechanism would operate if plant viruses move from cell to cell via the plasmodesmata. It is possible, as van Loon (1985) has suggested, that it is the fraction of PR proteins within the cell that have an antiviral role. It is also possible that some of the PR proteins may be active against other pathogens (such as bacteria and fungi) that are found in the intercellular spaces.

Only four of the PR proteins have been studied in any detail (PR proteins 1a, 1b, 1c and 2) and van Loon (1985) has shown that, in Samsun NN, these are not isozymes of 25 different enzymes known to increase in activity following TMV infection. Similar studies need to be made with the other PR proteins now recognized. Schlumbaum et al (1986) described the induction by pathogens in beans of a PR-like protein with chitinase activity which is a potent inhibitor of fungal growth. It would be interesting to know if any of the recognized PR proteins in tobacco have a similar activity (see also chapter by Fritig et al, this volume).

The evidence shows that PR proteins are closely associated with the localization of and induced resistance to viruses, but the exact nature of their action is not yet known. Because of the coordinated response of tobacco to different localized pathogens, it is possible that different PR proteins may be effective against different pathogens. The widespread occurrence of PR proteins in the angiosperms suggests that they may have an important role in the response of plants to pathogens.

References

Ahl P, Gianinazzi S 1982 b-Protein as a constitutive component in highly (TMV) resistant interspecific hybrids of *Nicotiana glutinosa* × *Nicotiana debneyi*. Plant Sci Lett 26:173–181

Ahl P, Antoniw JF, White RF, Gianinazzi S 1985 Biochemical and serological characterization of b-proteins from *Nicotiana* species. Plant Mol Biol 4:31–37

Antoniw JF, White RF 1983 Biochemical properties of the pathogenesis-related proteins from tobacco. Neth J Plant Pathol 89:255–264

Antoniw JF, White RF 1986 Changes with time in the distribution of virus and PR protein around single local lesions of TMV infected tobacco. Plant Mol Biol 6:145–149

Antoniw JF, Ritter CE, Pierpoint WS, van Loon LC 1980 Comparison of three pathogenesis-related proteins from plants of two cultivars of tobacco infected with TMV. J Gen Virol 47:79–87

Antoniw JF, White RF, Barbara DJ, Jones P, Longley A 1985 The detection of PR (b) protein and TMV by ELISA in systemic and localized virus infections of tobacco. Plant Mol Biol 4:55–60

Dumas E, Gianinazzi S 1986 Pathogenesis-related (b) proteins do not play a central role in TMV localization in *Nicotiana rustica*. Physiol Mol Plant Pathol 28:243–250

Fortin MG, Parent JG, Asselin A 1985 Comparative study of two groups of b proteins (pathogenesis related) from the intercellular fluid of *Nicotiana* leaf tissue infected by tobacco mosaic virus. Can J Bot 63:932–937

Fraser RSS 1981 Evidence for the occurrence of the 'pathogenesis-related' proteins in leaves of healthy tobacco plants during flowering. Physiol Plant Pathol 19:69–76

Fraser RSS 1982 Are 'pathogenesis-related' proteins involved in acquired systemic resistance of tobacco plants to tobacco mosaic virus? J Gen Virol 58:305–313

Fritig B, Kauffmann S, Dumas B, Geoffrey P, Kopp M, Legrand M 1987 Mechanism of the hypersensitivity reaction of plants. In: Plant resistance to viruses. Wiley, Chichester (Ciba Found Symp 133) p 92–108

Gianinazzi S 1983 Genetic and molecular aspects of resistance induced by infections or chemicals. In: Nester EW, Kosuge T (eds) Plant–microbe interactions: Molecular and genetic perspectives. Macmillan, New York, vol 1:321–342

Gianinazzi S, Martin C, Vallee JC 1970 Hypersensibilité aux virus, température et protéines solubles chez le *Nicotiana* Xanthi nc. Apparition de nouvelles macro-molécules lors de la répression de la synthèse virale. CR Hebd Séances Acad Sci Ser D Sci Nat 270:2383–2386

Jamet E, Fritig B 1986 Purification and characterization of 8 of the pathogenesis-related proteins in tobacco leaves reacting hypersensitively to tobacco mosaic virus. Plant Mol Biol 6:69–80

Nassuth A, Sänger HL 1986 Immunological relationship between 'pathogenesis-related' leaf proteins from tomato, tobacco and cowpea. Virus Res 4:229–242

Parent JG, Asselin A 1984 Detection of pathogenesis-related (PR or b) and other proteins in the intercellular fluid of hypersensitive plants infected with tobacco mosaic virus. Can J Bot 62:564–569

Redolfi P 1983 Occurrence of pathogenesis-related (b) and similar proteins in different species. Neth J Plant Pathol 89:245–254

Ross AF 1966 Systemic effects of local lesion formation. In: Beemster ABR, Dijkstra J (eds) Viruses of plants, North-Holland, Amsterdam, p 127–150

Schlumbaum A, Mauch F, Vogeli U, Boller T 1986 Plant chitinases are potent inhibitors of fungal growth. Nature (Lond) 324:365–367

van Loon LC 1977 Induction by 2-chloroethylphosphonic acid of viral-like lesions, associated proteins, and systemic resistance in tobacco. Virology 80:417–420

van Loon LC 1985 Pathogenesis-related proteins. Plant Mol Biol 4:111–116

van Loon LC, van Kammen A 1970 Polyacrylamide disc electrophoresis of the soluble leaf proteins from *Nicotiana tabacum* var 'Samsun' and 'Samsun NN'. II. Changes in protein constitution after infection with tobacco mosaic virus. Virology 40:199–211

Weststeijn EA 1981 Lesion growth and virus localization in leaves of *Nicotiana tabacum* cv. Xanthi-nc after inoculation with tobacco mosaic virus and incubation alternately at 22 °C and 32 °C. Physiol Plant Pathol 18:357–368

White RF 1983 Serological detection of pathogenesis-related proteins. Neth J Plant Pathol 89:311

White RF, Antoniw JF, Carr JP, Woods RD 1983 The effects of aspirin and polyacrylic acid on the multiplication and spread of TMV in different cultivars of tobacco with and without the N-gene. Phytopathol Z 107:224–232

White RF, Dumas E, Shaw P, Antoniw JF 1986 The chemical induction of PR (b) proteins and resistance to TMV infection in tobacco. Antiviral Res 6:177–185

White RF, Rybicki EP, Von Wechmar MB, Dekker JL, Antoniw JF 1987 Detection of PR1 type proteins in *Amaranthaceae, Chenopodiaceae, Gramineae* and *Solanaceae* by immunoelectroblotting. J Gen Virol 68:2043–2048

DISCUSSION

Dodds: In systemic induced resistance experiments with TMV against TMV (Ross 1961, McIntyre et al 1981), one can quantify where the maximum response occurs, in terms of which leaf positions, distant from the one original-ly induced, show the greatest reduction in the number or size of lesions. The actual leaf showing maximum response depends on the experimental conditions. This implies that PR protein levels should also be highest in that leaf: is there any evidence for that? And what is the evidence for high concentrations of PR protein in distant leaves? I am wondering if the tissue between lesions on an inoculated leaf shows the same capacity to reduce lesion size as tissue from distant leaves, where lesions also occur in small numbers and small sizes.

Antoniw: I don't know of any studies in which the levels of PR proteins and induced resistance were measured in different leaf positions relative to the inoculated leaf. The levels of PR proteins in distant systemically resistant leaves are not high; they reach only about 10% of the levels in the inoculated leaf.

Fraser: I compared the amounts of PR proteins in lower and upper leaves and measured the amount of resistance, either by lesion diameter or lesion size. There was a tenfold difference in protein concentration between lower and upper leaves, but similar amounts of resistance. This suggests that PR proteins are not quantitatively related to resistance. We have been doing these experiments, producing correlative evidence for and against a role for PR proteins in resistance, for the past decade: we are not going to resolve the problem that way. The way forward is going to be by using transgenic plants and asking whether they are resistant. The problem will also be resolved if PR proteins can be shown to have a particular enzyme activity.

Fritig: One should consider the level of response per cell. To find a correlation with any metabolic change one should be able to measure the intensity of the alteration. This is lacking at present.

Fraser: One should also measure the effect on virus multiplication.

Fritig: As I shall describe in my paper (Fritig et al, this volume), we have assigned functions to several of the PR proteins. Many of these proteins are hydrolases: some are chitinases, others are 1,3-β-glucanases. However, this still does not fully explain their role in the plant's response to viral infection.

Sela: Dr Antoniw, you say that the PR proteins are produced in the right place at the right time. Your TMV growth curve differs from any growth curve that I know, in either a localized or a systemic infection. Usually, after three to six hours, one starts to notice activity increase in both local and systemic infections. Then over the next 36–40 hours the systemic infection continues to increase but the localized infection stops. You say that TMV does not come up at all until after about four days.

Antoniw: Our data in Fig. 2 show the levels of PR-1a and TMV in systemic and localized infections of tobacco leaves at one day intervals. The earliest time point was 24 hours. We found small amounts of TMV at the earliest time points, but the largest increase in TMV occurred after day 4.

Sela: If we agree that the early stages of localization take place in the initial 36–42 hours, the PR proteins are late proteins. If they participate in defining localization they must be produced at an earlier time.

Antoniw: In the localized infection the largest amounts of PR-1 were produced after four days, at the time that the increase in virus was being restricted. The localization mechanism is not a fast process — in *N. tabacum* and *N. rustica* it takes 10–12 days before localization is complete.

Sela: Do you agree that the PR proteins appear later and that they do not participate in virus localization? Virus replication stops some time before the PR proteins are made.

Antoniw: No, I don't agree. Our data in Fig. 2 and elsewhere (Antoniw et al 1984, 1985) show that large amounts of PR proteins are produced before virus replication stops. It is therefore quite possible for PR proteins to be involved in virus localization.

Matthews: You showed a histogram arranged as a cross-section through a lesion, with PR-1 concentrated at the edge of the lesion and very little PR-1 in the central region (Fig. 3). Could that be an artefact, due to the cells in the centre dying, and therefore not synthesizing anything?

Antoniw: That is one possible explanation, but if it is a natural consequence of virus localization then it is not an artefact. By the time that the area of necrosis stopped growing it had covered the central 5 mm diameter disc (Fig. 3). However, the highest levels of virus were found in this central area. The suggestion that the distribution of PR-1 is associated with restriction of the virus is based on the observation that there are high levels of PR proteins at the edge of the lesion where the virus is being restricted.

Matthews: What is the effect of aspirin on virus replication where there is no necrosis, as in a chlorotic lesion? Does aspirin itself have antiviral activity?

Antoniw: In systemic infections of, for example, Xanthi or Samsun, which do not have the *N* gene, PR proteins are induced by aspirin. Aspirin reduces virus multiplication and spread, but in our hands it does not convert a systemic infection into a localized one. However, it is possible that aspirin does not induce high enough levels of PR proteins.

Loebenstein: What is the situation in a chlorotic lesion and in a starch lesion? I am wondering about PR proteins in systems where there are living cells, as in chlorotic lesions, and virus is still localized. Secondly, when you apply chemicals to induce resistance, in cases where there is no systemic effect, are PR proteins found in the upper leaves?

Antoniw: We haven't looked at the distribution of PR proteins across chlorotic lesions. Most chemical inducers, such as polyacrylic acid, induce PR proteins and resistance only in the treated leaf. Others, like ethephon, induce both PR proteins and resistance in the treated leaf and other leaves on the same plant. With salicylic acid, PR proteins were induced only in the treated leaf; resistance was induced in the treated leaf but also occasionally in upper untreated leaves (van Loon & Antoniw 1982).

Zaitlin: Did you say that the lesion collapses and the virus cannot leave the collapsed area?

Antoniw: Weststeijn (1981) did temperature shift experiments to look at the ability of TMV to escape from local necrotic lesions of different ages. In fully mature, 12-day-old lesions, the virus was completely localized.

Zaitlin: But not necessarily trapped within the necrotic area? Your own results suggest that there is virus *outside* the necrotic area. That has also been shown with electron microscopy. One can even cut tissue from an area outside the lesion and demonstrate infectious virus there.

Antoniw: Has that been done with fully mature lesions?

Loebenstein: Bob Milne showed by electron microscopy that in the living cells around the chlorotic area there is virus and it does not spread (Milne 1966).

Gianinazzi: That is true; in fact when the plant is shifted from 20°C to 32°C the virus multiplication starts again and the virus spreads throughout the plant (Kassanis 1952). However, I observed that if the transfer to 32°C is made three weeks after the appearance of the necrotic local lesions, spread of the virus no longer occurs (unpublished results).

Zaitlin: That is not necessarily due to the necrotic lesion itself. There is something restricting the virus in that area, but it is not the necrosis that restricts it. The observation by Dr Antoniw was that virus was restricted from moving because it was in the necrotic tissue. In the cirumstances you described, it is outside the necrotic tissue and still cannot move.

Fritig: We injected [³H]uridine around necrotic lesions to distinguish between the virus which was already present and the virus which was newly made. There is definitely [³H]uridine going into the virus in the zone of living cells around the necrotic lesion: therefore there is virus replication for at least seven days after inoculation but it is at a low level. The localizing mechanism is not clear-cut, but acts gradually across the tissue.

Sela: When did you give the first injections of [³H]uridine?

Fritig: We injected a few hours before harvesting the tissue, seven days after inoculation. We have also done this at very early stages of the infection: before, at and shortly after the onset of necrotization (Fritig et al, this volume).

References

Antoniw JF, White RF, Barbara DJ, Longley A 1984 Virus and pathogenesis-related protein accumulation after tobacco mosaic virus infection of tobacco. Biochem Soc Trans 12:828

Antoniw JF, White RF, Barbara DJ, Jones P, Longley A 1985 The detection of PR (b) proteins and TMV by ELISA in systemic and localized virus infections of tobacco. Plant Mol Biol 4:55–60

Fritig B, Kauffmann S, Dumas B, Geoffroy P, Kopp M, Legrand M 1987 Mechanism of the hypersensitivity reaction of plants. In: Plant resistance to viruses. Wiley, Chichester (Ciba Found Symp 133) p 92–108

Kassanis B 1952 Some effects of high temperature on the susceptibility of plants to infection with viruses. Ann Appl Biol 39:358–369

McIntyre JL, Dodds JA, Hare JD 1981 Effects of localized infections of *Nicotiana tabacum* by tobacco mosaic virus on systemic resistance against diverse pathogens and an insect. Phytopathology 71:297–301

Milne RG 1966 Electron microscopy of tobacco mosaic virus in leaves of *Nicotiana glutinosa*. Virology 28:527–532

Ross AF 1961 Systemic acquired resistance induced by localised virus infection in plants. Virology 14:340–358

Van Loon LC, Antoniw JF 1982 Comparison of the effects of salicylic acid and ethephon with virus-induced hypersensitivity and acquired resistance in tobacco. Neth J Plant Pathol 88:237–256

Weststeijn EA 1981 Lesion growth and virus localization in leaves of *N.tabacum* cv. Xanthi-nc after inoculation with tobacco mosaic virus and incubation alternately at 22°C and 32°C. Physiol Plant Pathol 18:357–368

Characterization of pathogenesis-related proteins and genes

J.F. Bol, R.A.M. Hooft van Huijsduijnen*, B.J.C. Cornelissen† and J.A.L. van Kan

Department of Biochemistry, Leiden University, P.O. Box 9505, 2300 RA Leiden, The Netherlands

Abstract. cDNA clones corresponding to seven classes of mRNAs (clusters A to G) that are induced by tobacco mosaic virus (TMV) infection of Samsun NN tobacco were used to characterize the corresponding proteins and genes. cDNAs were found to correspond with the known acidic pathogenesis-related (PR) proteins 1a, 1b and 1c (cluster B), P and Q (cluster D) and probably R or S (cluster E). Clusters G and F correspond to basic proteins homologous to those encoded by clusters B and D, respectively. The observation that clusters D and F correspond to tobacco chitinases may partly explain the TMV-induced resistance to chitin-containing pathogens. The number of genes in the Samsun NN genome corresponding to clusters A, B, C and G was found to be higher than the number corresponding to clusters D, E and F.

1987 Plant resistance to viruses. Wiley, Chichester (Ciba Foundation Symposium 133) p 72–91

Infection with viruses, viroids, bacteria or fungi as well as treatment with certain chemicals is known to induce the synthesis of pathogenesis-related (PR) proteins in at least 16 plant species (van Loon 1985). These proteins are characterized by their solubility at low pH, their accumulation in the intercellular space of the leaf (Parent & Asselin 1984), their resistance to proteases, and their migration in alkaline polyacrylamide gels because of their acidic nature (van Loon 1982). Ten PR proteins, induced by tobacco mosaic virus (TMV) infection of the tobacco cultivars Samsun NN or Xanthi nc, have been studied in most detail. These are the PR proteins 1a, 1b, 1c, 2, N, O, P, Q, R and S, which, with the exception of R and S, have been purified to homogeneity (Jamet & Fritig 1986). The PR-1 proteins (molecular mass ≈ 15 kDa) are serologically related (Hooft van Huijsduijnen et al 1985) and so are PR proteins 2, N and O (molecular mass ≈ 40 kDa) (Fortin et al 1985). Recently, P and Q (molecular mass ≈ 27 kDa) were found to constitute a

---

* *Present address*: Laboratoire de Génétique Moléculaire des Eucaryotes, 11 Rue Human, 67085 Strasbourg, France.
† *Present address*: MOGEN International B.V., Einsteinweg 97, 2333 CB Leiden, The Netherlands.

third class of serologically related proteins (Hooft van Huijsduijnen et al, 1987).

PR proteins have attracted considerable interest because their occurrence is closely associated with the induction of a broad response of the plant to (further) infection with viruses, fungi or bacteria (van Loon 1982, Kuc 1982). As a first step towards an understanding of the structure and function of PR proteins and their mode of induction by pathogens, we have cloned DNA copies of a number of mRNAs that are induced by TMV infection of Samsun NN tobacco. These cDNAs have been used to identify corresponding proteins by hybrid-selected translation, to analyse genes for PR proteins on genomic blots and to isolate clones of PR genes from a genomic library of Samsun NN tobacco. Here we shall present some of the results, together with evidence on the possible role of PR proteins in defence responses of the plant to viruses and other pathogens.

Induction of PR proteins by virus infection and salicylic acid treatment

In addition to the synthesis of PR proteins, infection of tobacco plants with TMV results in an increase in the activity of at least 25 enzymes (Antoniw & White 1983). However, spraying of tobacco with salicylic acid selectively induces a subset of PR proteins as well as resistance to TMV infection (White 1979). Fig. 1 shows the proteins in the intercellular fluid from cowpea, bean and tobacco plants that have been sprayed with salicylic acid (lanes S) or infected with viruses (lanes A and T), compared with healthy controls (lanes H). In tobacco, TMV induces necrotic lesions and a number of PR proteins (lane 13) which are absent in the healthy plant (lane 9). A subset of these PR proteins, notably 1a, 1b, 2 and N, is also induced by salicylic acid (lane 11). Alfalfa mosaic virus (AlMV) replicates systemically in tobacco and does not induce the synthesis of PR proteins (lane 12). However, if plants are sprayed with salicylic acid and are subsequently inoculated with AlMV, virus multiplication is inhibited by 90% (Hooft van Huijsduijnen et al 1986a). In bean plants AlMV induces necrotic lesions and the synthesis of several proteins (lane 8), one of which is also induced by salicylic acid (lane 7). Pretreatment of the plants with salicylic acid reduced the number of AlMV lesions by 75%. In cowpea plants infection with AlMV (lane 4) and spraying with salicylic acid (lane 3) both induce the synthesis of at least one protein that is absent in healthy plants (lane 1). Salicylic acid was found to inhibit AlMV replication in cowpea protoplasts by up to 99% by a specific block of viral RNA replication without any detectable inhibition of host metabolism (Hooft van Huijsduijnen et al 1986a). Spraying of the plants with another aromatic component, p-coumaric acid, did not induce the synthesis of PR proteins (Fig. 1, lanes C) and did not inhibit virus multiplication. Possibly, the salicylic acid-induced PR proteins play a role in the antiviral response.

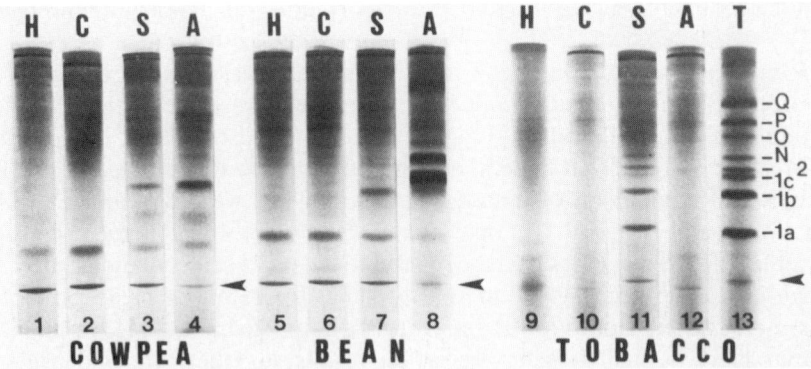

FIG. 1. Accumulation of PR proteins in the intercellular fluid of cowpea (lanes 1 to 4), bean (lanes 5 to 8) and tobacco (lanes 9 to 13) after various treatments. Plants were sprayed with water (lanes H), *p*-coumaric acid (lanes C), or salicylic acid (lanes S), or inoculated with AlMV (lanes A) or TMV (lane T). Samples of the intercellular fluid were electrophoresed in non-denaturing polyacrylamide gels; the position of the bromophenol blue marker is indicated by arrows. The nomenclature of the tobacco PR proteins is according to van Loon (1982). (Reproduced by permission from Hooft van Huijsduijnen et al 1986a; experimental details are given in this reference.)

Analysis of cDNA clones to PR mRNAs

Screening a cDNA library to poly(A)$^+$ RNA from TMV-infected tobacco by a differential hybridization procedure allowed us to isolate cDNA clones corresponding to six classes of TMV-inducible mRNAs, designated clusters A to F (Hooft van Huijsduijnen et al 1986b). The Northern blots shown in Fig. 2 demonstrate that the mRNAs occur at a low level in healthy plants (lanes H) but are strongly induced in TMV-infected plants (lanes T). Some of the mRNAs, notably B and C, are also strongly induced by spraying tobacco with salicylic acid (lanes S). Hybrid-selected translation with cluster B, D, E and F cDNAs yielded proteins that were immunoprecipitated with an antiserum to a mixture of PR proteins (Hooft van Huijsduijnen et al 1986b).

The sequences of almost all the available cDNA clones have been determined. Cluster B was found to correspond to the PR-1 proteins (Cornelissen et al 1986a). However, the situation turned out to be more complex than was anticipated from the known pattern of PR proteins. cDNAs were identified to three mRNAs encoding acidic PR-1 proteins with more than 90% amino acid sequence homology (PR-1a, -1b and -1c). In addition, cDNA clones were identified corresponding to a more basic PR protein that showed a 67% homology to the acidic PR-1 proteins. This protein exceeds the acidic PR-1 proteins in molecular weight because of a C-terminal extension of 36 amino acids (B.J.C. Cornelissen et al, unpublished).

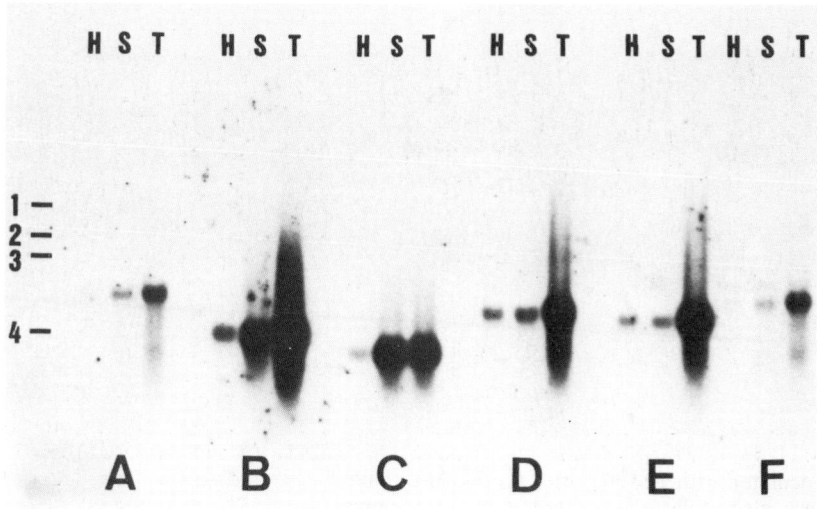

FIG. 2. Induction of mRNAs A to F in tobacco by salicylic acid spraying and TMV infection. A Northern blot of 20 μg poly(A)$^+$ RNA from healthy tobacco leaves (lanes H), from leaves sprayed with salicylic acid (lanes S) and from TMV-infected leaves (lanes T) was cut into six strips. These were hybridized separately to ^{32}P-labelled plasmids from clones corresponding to clusters A to F. The strips were realigned in their original orientation for autoradiography. The positions of AlMV RNAs 1 (3644 nucleotides), 2 (2593 nt), 3 (2037 nt) and 4 (881 nt) are indicated on the left. (Reproduced by permission from Hooft van Huijsduijnen et al 1986b; experimental details are given in this reference.)

Recently, antisera to purified PR-P and -Q were used to demonstrate a close serological relationship between these two PR proteins (Hooft van Huijsduijnen et al 1987). The antisera gave a strong reaction with the ≈ 25 kDa protein translated from hybrid-selected cluster D mRNA and a weak reaction with the ≈ 34 kDa protein translated from cluster F mRNA. Although the cDNA clones are incomplete, the data indicate a 65% amino acid sequence homology between the cluser D-encoded acidic PR proteins P and Q and the cluster F-encoded basic protein(s) (Hooft van Huijsduijnen et al 1987).

A nearly full-length cluster E cDNA clone was found to encode a protein of 226 amino acids with a putative signal sequence of 25 amino acids (Cornelissen et al 1986b). The amino acid composition and relative molecular mass of this mature protein are very similar to those determined for the purified PR protein R by Pierpoint (1986). Antisera to PR-R, prepared by B. Fritig and co-workers, are now being tested in our laboratory to confirm that cluster E corresponds to PR-R and possibly related PR proteins.

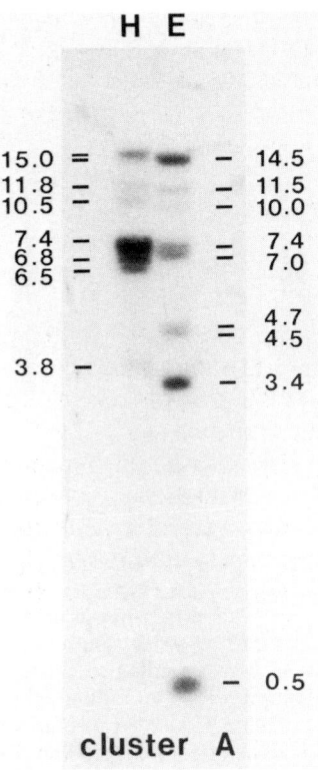

FIG. 3.   Southern blot analysis of cluster A genes in genomic DNA of Samsun NN
tobacco. A filter containing genomic DNA digested with HindIII (lane H) or
EcoRI (lane E) was hybridized to the nick-translated insert (405 bp) excised with
HindIII and EcoRI from a pUC9 vector containing cluster A cDNA (PROB 40).
Experimental details are described elsewhere (Hooft van Huijsduijnen et al 1987). The
estimated size of the fragments (kb) is indicated.

Analysis of PR genes

To obtain insight into the complexity of the gene families corresponding to
the TMV-inducible mRNAs shown in Fig. 2, we have hybridized the corresponding cDNAs to Samsun NN genomic DNA previously digested with
HindIII or EcoRI. The results with the cluster A probe are shown in Fig. 3.
The HindIII and EcoRI digests of tobacco DNA both gave a complex pattern
of eight to nine bands of unequal intensity, suggesting that cluster A corresponds to a multigene family. With the cluster B probe, corresponding to
acidic PR-1 proteins, eight bands were detectable with the two digests of
tobacco DNA. Genomic clones corresponding to three of these bands were
isolated from a genomic library and were found each to contain a single PR-1
gene (B.J.C. Cornelissen et al, unpublished results). This suggests the pres-

ence of eight genes for acidic PR-1 genes in the Samsun NN genome. Under stringent conditions the cDNA corresponding to the basic PR-1 protein did not cross-hybridize with the genes for the acidic homologues. Instead, this probe hybridized to different sets of fragments of the digested tobacco DNA, suggesting the presence of approximately eight genes for basic PR-1 proteins in the Samsun NN genome (B.J.C. Cornelissen et al, unpublished results). Therefore, the genes for basic PR-1 proteins are considered to represent a separate cluster of TMV-inducible genes, designated cluster G.

The complexity of the hybridization pattern obtained with cluster C cDNA probes was similar to that of the clusters A, B and G patterns. Four different genomic clones were isolated, two of which were sequenced (J.A.L. van Kan et al, unpublished). In contrast to the cluster B genes, the cluster C genes were found to contain introns. These genes encode polypeptides of 109 amino acids with a high proportion of glycine and charged amino acids.

Compared to clusters A, B, C and G, the genomic blots of clusters D, E and F showed a simpler pattern. With probes corresponding to the 3' terminal regions of the genes, two bands were obtained with the EcoRI digests for each cluster (Hooft van Huijsduijnen et al 1987). This indicates that clusters D, E and F are each represented by two to four genes in the amphidiploid genome of Samsun NN tobacco.

Possible functions of PR proteins

The observation that genes which are salicylic acid-inducible in tobacco plants are also rapidly expressed after adding salicylic acid to protoplasts (R.A.M. Hooft van Huijsduijnen, unpublished results) supports the hypothesis that proteins encoded by these genes play a role in the salicylic acid-mediated inhibition of viral RNA synthesis in these protoplasts. On the other hand, an intracellular function is difficult to reconcile with the fact that PR proteins are known to be excreted by the plant cells. The translation product of mRNA C, which is most strongly induced by salicylic acid, is not precipitated by an antiserum to PR proteins (Hooft van Huijsduijnen et al 1986b). Although the N-terminus of this protein has the characteristics of a signal peptide, the possibility exists that it accumulates intracellularly. In this case, it could be the putative antiviral factor.

Recently, the sequences of basic chitinases from bean (Broglie et al 1986) and tobacco (Shinshi et al 1987) have been published. Our cluster F sequence is identical to that of the tobacco chitinase and 76% homologous to that of the bean chitinase (Hooft van Huijsduijnen et al 1987). Two acidic chitinases corresponding to PR-P and -Q, and two basic chitinases probably corresponding to cluster F, have been identified in TMV-infected tobacco by M. Legrand et al (unpublished results). As chitinases are known to be potent inhibitors of bacterial and fungal growth (Schlumbaum et al 1986 and references therein), the PR proteins corresponding to clusters D and F may play a role in the TMV-induced resistance of tobacco against chitin-containing pathogens. An enzyme hydrolysing the other major component of the cell wall of these pathogens, β-1,3-glucan, is known to be induced by TMV infection of *Nicotiana glutinosa* (Moore & Stone 1972).

The cluster E-encoded PR protein shows a 70% homology to thaumatin, the intensely sweet-tasting protein that accumulates in the fruits of *Thaumatococcus danielii* Benth., a West African rain forest shrub (Cornelissen et al 1986b). It will be interesting to see whether this protein also has a specialized function as part of a plant defence mechanism.

Several characteristics of TMV-induced mRNAs and their encoded proteins are summarized in Table 1. A possible role of PR proteins in the acquired systemic resistance of plants has often been discussed (for references see Hooft van Huijsduijnen et al 1986a). The finding that some of these proteins correspond to hydrolytic enzymes capable of degrading the cell walls of chitinous pathogens lends support to the view that different groups of PR proteins are designed to combat different types of pests. If PR proteins are also involved in antiviral responses, the salicylic acid-inducible proteins are the most likely candidates for this role. In principle, a possible role for PR proteins in a resistance mechanism could be investigated by the construction of transgenic plants producing one or more of these proteins constitutively. However, the finding that some of these proteins are encoded by rather complex gene families may complicate this type of experiment. Within one family the genes seem to be 90% homologous but small differences may affect the function of the encoded proteins. In transformation experiments, preference should be given to genes that are known to be expressed because the corresponding cDNA clones have been identified. Although the physiological meaning of the occurrence of acidic and basic versions of several PR proteins is not yet known, transformation experiments should include both types of

TABLE 1   Characteristics of TMV-induced mRNAs and proteins

Cluster	Size of hybrid-selected translation product[a]	Induction by salicylic acid	Corresponding protein
A	40	+	?
B	15	+ +	Acidic PR-1 proteins
C	15	+ + +	Hydrophilic protein of 109 amino acids
D	25	−	Acidic chitinases, PR-P/Q
E	25	−	PR-R?, homologous to thaumatin
F	34	±	Basic chitinases
G	?	?	Basic PR-1 proteins

[a] Apparent molecular mass in kDa; data are from Hooft van Huijsduijnen et al (1986b).

genes because of the possibility that a coordinated action of these proteins is required.

Acknowledgement

The assistance of Mr F.Th. Brederode in the experimental work is gratefully acknowledged.

References

Antoniw JF, White RF 1983 Biochemical properties of the pathogenesis-related proteins from tobacco. Neth J Plant Pathol 89:255–264

Broglie KE, Gaynor JJ, Broglie RM 1986 Ethylene-regulated gene expression: molecular cloning of the genes encoding an endochitinase from *Phaseolus vulgaris*. Proc Natl Acad Sci USA 83:6820–6824

Cornelissen BJC, Hooft van Huijsduijnen RAM, van Loon LC, Bol JF 1986a Molecular characterization of the messenger RNAs for 'pathogenesis-related' proteins 1a, 1b and 1c, induced by TMV infection of tobacco. EMBO (Eur Mol Biol Organ) J 5:37–40

Cornelissen BJC, Hooft van Huijsduijnen RAM, Bol JF 1986b A tobacco mosaic virus-induced tobacco protein is homologous to the sweet-tasting protein thaumatin. Nature (Lond) 321:531–532

Fortin MG, Parent JG, Asselin A 1985 Comparative study of two groups of b proteins (pathogenesis related) from the intercellular fluid of *Nicotiana* leaf tissue infected by tobacco mosaic virus. Can J Bot 63:932–937

Hooft van Huijsduijnen RAM, Cornelissen BJC, van Loon LC, van Boom JH, Tromp M, Bol JF 1985 Virus-induced synthesis of pathogenesis-related proteins in tobacco. EMBO (Eur Mol Biol Organ) J 4:2167–2171

Hooft van Huijsduijnen RAM, Alblas SW, De Rijk RH, Bol JF 1986a Induction by salicylic acid of pathogenesis-related proteins and resistance to alfalfa mosaic virus infection in various plant species. J Gen Virol 67:2135–2143

Hooft van Huijsduijnen RAM, van Loon LC, Bol JF 1986b cDNA cloning of six mRNAs induced by TMV infection of tobacco and a characterization of their translation products. EMBO (Eur Mol Biol Organ) J 5:2057–2061

Hooft van Huijsduijnen RAM, Kauffmann S, Brederode F.Th, Cornelissen BJC, Legrand M, Fritig B, Bol JF 1987 Homology between chitinases that are induced by TMV infection of tobacco. Plant Mol Biol, in press

Jamet E, Fritig B 1986 Purification and characterization of 8 of the pathogenesis-related proteins in tobacco leaves reacting hypersensitively to tobacco mosaic virus. Plant Mol Biol 6:69–80

Kuc J 1982 Induced immunity to plant disease. Bioscience 32:854–860

Moore AE, Stone BA 1972 Effect of infection with TMV and other viruses on the level of a β-1,3-glucan hydrolase in leaves of *Nicotiana glutinosa*. Virology 50:791–798

Parent JG, Asselin A 1984 Detection of pathogenesis-related proteins (PR or b) and of other proteins in the intercellular fluid of hypersensitive plants infected with tobacco mosaic virus. Can J Bot 62:564–569

Pierpoint WS 1986 The pathogenesis-related proteins of tobacco leaves. Phytochemistry (Oxf) 25:1595–1601

Schlumbaum A, Mauch F, Vogeli U, Boller T 1986 Plant chitinases are potent inhibitors of fungal growth. Nature (Lond) 324:365–367

Shinshi H, Mohnen D, Meins F 1987 Regulation of a plant pathogenesis-related enzyme: inhibition of chitinase and chitinase mRNA accumulation in cultured tobacco tissues by auxin and cytokinin. Proc Natl Acad Sci USA 84:89–93

van Loon LC 1982 Regulations of changes in proteins and enzymes associated with the active defence against virus infection. In: Wood RKS (ed) Active defence mechanisms in plants. Plenum, New York, p 247–274

van Loon LC 1985 Pathogenesis-related proteins. Plant Mol Biol 4:111–116

White RF 1979 Acetylsalicylic acid (aspirin) induces resistance to tobacco mosaic virus in tobacco. Virology 99:410–412

DISCUSSION

Beachy: Dr Bol, is there no nucleotide homology 5′ to the CAAT boxes of the promoters that have been sequenced?

Bol: There is some homology immediately upstream of the Cap sites of the cluster B and C genes, but no clear homology is detectable upstream of the CAAT boxes.

Beachy: Have you also made chimeric genes between the promoters and marker genes?

Bol: Yes, we have fused the promoter regions to the CAT gene and we are now introducing these constructs into protoplasts by electroporation, to see if the CAT gene is inducible by salicylic acid.

Duffus: Do PR proteins occur during limited systemic infections that involve necrotic lesions which escape to another part of the plant? Beet necrotic yellow vein and tomato bushy stunt are examples of viruses that behave like this.

Bol: If we infect salicylic acid-treated plants with a virus that normally gives a systemic infection without a necrotic reaction, there is an almost complete block of virus multiplication. I think that if salicylic acid-induced PR proteins are involved in that inhibition, it will occur in the primary inoculated cells because it also occurs in protoplasts. The necrotic reaction is just the trigger that leads to the production of ethylene, which itself triggers the production of all the PR proteins and a number of other enzymes. The inhibition of virus multiplication is not related to the necrotic reaction.

Fraser: Dr Bol, what evidence shows that salicylic acid inhibits TMV multiplication in protoplasts by inducing proteins rather than by direct antiviral action?

Bol: In the protoplast, salicylic acid completely inhibits production of viral plus-strand and minus-strand RNAs. This suggests that it acts at the level of the production of double-stranded RNA. We added salicylic acid to an enzyme system that specifically produces alfalfa mosaic virus RNA *in vitro* and it had no inhibitory effect. Therefore I suggest that salicylic acid does not directly inhibit the replicase, although the *in vitro* study only investigates a limited number of enzymic steps. We have not yet added purified PR proteins to that replicase system to see what effect they have. The PR-1 proteins are extracellular, so it is difficult to reconcile their action with an effect inside the cell. I still hope that the cluster C protein, although it has a signal peptide, may accumulate inside the cell and may be a good candidate for that antiviral function. We have to wait for the antiserum to see where the protein is located.

Gianinazzi: We recently localized the PR-b$_1$ (PR-1a) protein at the ultrastructural level. We did this in a PAA$^+$ line of tobacco which responds to

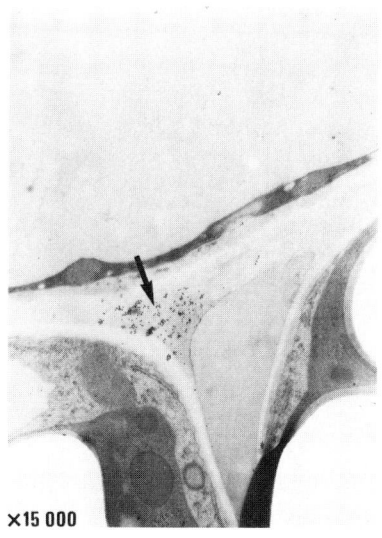
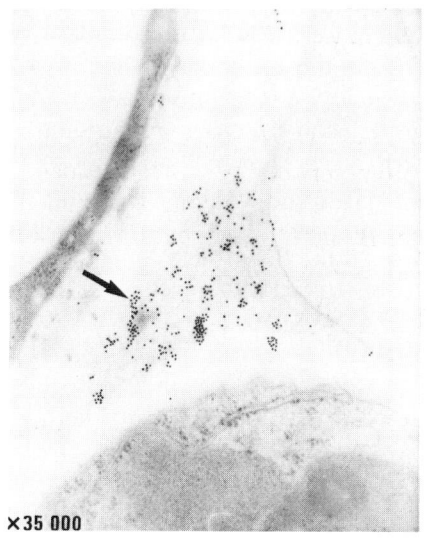

×15 000   ×35 000

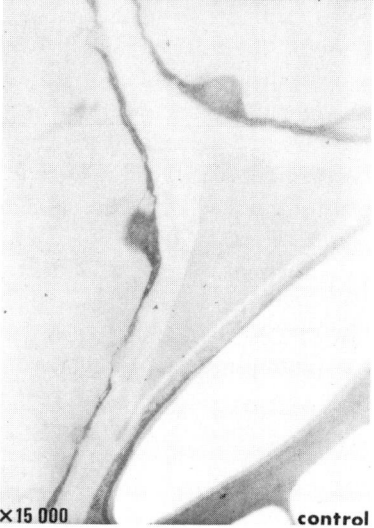

×15 000   control

FIG. 1.
(*Gianinazzi*). Immunocytochemical localization of PR-b$_1$ protein visualized using immuno-gold labelling ($\rightarrow$) (E. Dumas, J. Lherminier & S. Gianinazzi, unpublished results).

polyacrylic acid by producing PR-b proteins (Gianinazzi & Kassanis 1974, Dumas et al 1987). Using an antiserum to PR-b$_1$ provided by J.F. Antoniw and R.F. White we demonstrated that this protein accumulates in the middle lamella of leaves treated with polyacrylic acid (see Fig. 1). As protein PR-b$_1$ is serologically related to b$_1'$, b$_2$ and b$_3$ in this tobacco line, we probably detected these other proteins as well.

Davies: Do the proteins escape from protoplasts, Dr Bol?

Bol: The addition of salicylic acid to protoplasts results in the rapid induction of the mRNAs. Therefore, it is possible that their translation products also have an early effect on virus multiplication. The basic level of PR protein mRNA in freshly isolated protoplasts is very low. We have not analysed the proteins in the extracellular fluid.

Antoniw: We made protoplasts from induced leaves and found most of the PR proteins in the enzyme solution used to digest away the cell walls.

Harrison: Presumably this is not simply because the protoplasts were damaged by the isolation procedure.

Davies: Which proteins, in addition to protein C, have a signal sequence?

Bol: All the PR proteins analysed so far have a signal sequence: the PR-1 proteins, the cluster C protein, the thaumatin-like protein (cluster E) and the basic chitinase (cluster F).

Fritig: PR proteins are definitely induced by osmotic shock. It is therefore not surprising that, even in healthy protoplasts, production of PR proteins is observed. Just taking leaf tissue and infiltrating these with water by vacuum suction induces PR proteins.

Bol: We have observed that PR proteins are not induced by a heat shock.

Gianinazzi: A nutrient deficiency shock also induces PR-b proteins. We grew *Nicotiana tabacum* plants *in vitro* with a low level of nutrients and PR proteins were induced. When we transferred the plants to a nutrient-rich medium, still *in vitro*, these proteins disappeared (Ahl & Gianinazzi 1982).

Beachy: What is the time course of turnover of these proteins, Dr Bol?

Bol: We have only analysed the mRNAs. We inoculated the lower leaves of Samsun NN tobacco with TMV. From the second day we started to detect messenger RNAs. By eight days all six mRNAs are induced in the upper leaves. In the lower leaves, the larger ones decline after five days whereas the smaller ones are stable. We have not yet analysed the proteins.

Beachy: What is the time course if you spray the leaf with salicylic acid?

Bol: We have not studied the time course after spraying. However, two days after spraying, the level of mRNAs induced by salicylic acid is similar to that induced by TMV infections. Therefore spraying rapidly induces proteins in the intact plant as well as in protoplasts.

van Kammen: Are none of the PR proteins synthesized in isolated protoplasts?

White: Using the antiserum to PR-1, we showed that PR-1 is induced by salicylic acid in protoplasts and secreted into the culture medium. However, the amount of protein found was much less than the amounts produced in intact leaves. This may be partly due to the suppressive effect of mannitol on protein synthesis in the protoplasts.

Bol: I don't think TMV infection will induce these proteins in protoplasts. Ethylene or ethephon may induce the higher molecular weight PR proteins, but we have not studied that in protoplasts.

White: I would recommend caution in stating that PR proteins are stress proteins. One can stress tobacco plants in many ways and get no PR protein production. In the case of osmotic effects, mannitol, for instance, will induce PR protein production, but we have found that when mannitol is used together with antibiotics, no PR proteins are produced. In some cases, secondary bacterial and fungal infections might have induced PR proteins, so it may not be obvious which particular form of stress induced the proteins.

Harrison: By analogy, there are many ways of stopping a car from functioning properly! Ron Fraser found that PR proteins were produced at flowering time in apparently healthy tobacco plants. Are you suggesting that this was because the conditions in his glasshouse were not aseptic?

White: It might be favourable for plants to produce PR proteins at flowering time (in order to produce virus-free seed, for example). We found that certain plant growth substances will induce these proteins. Therefore it is reasonable to suppose that hormonal changes that occur during flowering are responsible for the induction.

Harrison: So your point is not that PR proteins are always related to a pathogen, but that they are not an invariable response to stress.

Fraser: Many things induce PR proteins, but the one thing that does not seem to induce them is plant viruses! Ninety-nine per cent of the virus infections which induce PR proteins are the necrotic ones—it is the necrosis that induces the PR proteins, not the viruses. In systemic infections there are very few cases of induction of PR proteins. Perhaps these few exceptional cases were actually responding to ethylene produced by necrotically responding plants in the same experiment. Generally, viruses giving systemic infections do not induce large amounts of PR proteins, so I would suggest that viruses are not relevant to PR proteins, or vice versa!

Sänger: We have looked at viroid infections and PR proteins. We studied and sequenced P14 (which is related to PR-1a from tobacco) from tomato leaves (Lucas et al 1985). Potato spindle tuber viroid (PSTV) infection in tomato is not strictly necrotic, like TMV in Xanthi nc, but we nevertheless observed a marked increase in P14. This also happens in other viroid–host combinations, such as citrus exocortis virus and *Gynura aurantiaca*. On the other hand, we have a continuously PSTV-synthesizing suspension culture of potato cells in which large amounts of PSTV are produced but not excreted. However, no PR proteins are found in these cells.

Beachy: How long after viroid infection of tomato plants do you make these studies? Is it at a late stage of infection, when the plants are heavily damaged?

Sänger: The PR proteins start to appear about three weeks after inoculation which is a rather late response but the reaction of the tomato plants is not necrotic in the strict sense.

Beachy: This means that necrosis is not a precondition for the induction of PR proteins, but unhealthy plants are necessary.

Sänger: Dr Fraser suggested that necrosis is the precondition for PR proteins.

Fraser: It is a good suggestion for viruses, but it does not have to apply to other inducers.

Sänger: The viroid-producing cells in suspension culture do not show any cytopathic effect, and, as I said, there is no PR protein synthesis.

Loebenstein: Dr Sänger, have you looked at leakage in such cells?

Sänger: Yes, but we could not find PR proteins in the medium. However, we have not looked for messenger RNAs for the PR proteins in these cells.

Loebenstein: Have you looked for electrolyte leakage in viroid-infected cells? Maybe necrosis causes an increase in electrolyte leakage, whereas in your tomato–viroid system there is an increase in membrane instability which causes a leakage that results in the formation of PR proteins.

Sänger: We have not done such studies. Would you not expect PR proteins to leak out under those conditions?

Loebenstein: I am considering a kind of disorganization or decompartmentalization of the cells and the induction of PR proteins as a result of this.

Beachy: Has anyone studied the infection of a systemic host by one of the severe strains of TMV which cause rosetting, short internodes and shoestring in leaves, but not necrosis?

Gianinazzi: I recently inoculated *N. tabacum* cv Samsun (nn) with a mixture of TMV and PVX. This induced many necrotic lesions, both in the inoculated leaves and in the leaves above them, and the plant was full of PR-b proteins, in spite of the fact that the infection became systemic. After a few weeks a new lateral shoot developed that was symptom free. I don't know how to interpret this result with regard to PR-b proteins, because on the one hand the virus spread systemically, whereas a few weeks later a new shoot without symptoms appeared.

Loebenstein: This kind of recovery is well known in many plants. In the prunus necrotic ring spot group, new shoots appear which do not show symptoms (Nyland et al 1976). I doubt that in a plant like this, where the virus is very unevenly distributed, the recovery is related to PR proteins.

Antoniw: Dr Gianinazzi, did you find any virus in the new shoot?

Gianinazzi: Yes, there was a small amount of virus.

White: We have found (Kassanis et al 1974) that PR proteins are produced during systemic virus infections but at lower concentrations than in localized infections. However, systemic infections induced similar amounts of PR protein and a similar form of resistance to some chemical treatments. Possibly it is the very high concentration of PR protein around a localized infection that prevents the systemic spread of the virus. Necrosis is not necessary for PR protein induction. Some systemic infections and some chemical treatments induce PR proteins without necrosis. On the other hand, necrosis induced by mechanical damage or certain chemicals does not give rise to PR proteins.

Sänger: Is it correct that the necrotic response of plants is always tissue mediated? Are there cases where the response occurs in virus-infected protoplasts or in virus-infected cells in suspension culture? I know of no such plant system (in contrast to virus-infected animal and bacterial cells where cytopathic effects and lysis are well-known phenomena).

Loebenstein: Weststeijn (1984) obtained a necrosis-inducing factor from infected Xanthi nc plants, added it to a protoplast solution and reported necrosis of the protoplasts.

Sänger: An explanation might be that necrosis only develops in intact plant tissue. This necrotic reaction induces the synthesis of the PR proteins. Then the occurrence of the PR proteins would also be tissue mediated or dependent.

Matthews: Has anybody tried to develop a system for studying this question and the question of the role of the cell wall? If one used appropriate enzymic digestions to make three types of suspension from the same batch of tissue— clumps of cells averaging 10 cells in size, single cells and protoplasts—one could do the induction experiments and investigate what difference the state of the cell makes.

Bol: We are studying this question by the opposite approach. We are looking at the regulatory elements of the genes, seeing which proteins are binding to the promoter, and then what metabolite is binding to that protein. When we inoculate the lower leaves of a tobacco plant the genes are also all induced in the upper leaves. Therefore some mobile compound is travelling from the lower to the upper leaves and activating those genes. There are two ways of analysing that: either one can look at the genes, as we are doing, or one can try to identify the metabolite moving through the plant.

Davies: In the experiment that Dr Matthews suggests where a comparison is made between protoplasts, partially de-walled cells and groups of cells, it would be difficult to introduce virus or RNA into cells. An efficient way of achieving this, whether there is a cell wall or not, is by electroporation. But does electroporation itself induce PR proteins?

Bol: Not in our hands! We have done that and it does not induce PR proteins.

Loebenstein: Dr Bol, in your protoplast experiments with salicylic acid in which you did not obtain replication of alfalfa mosaic virus, did you find either PR proteins or their messenger RNAs?

Bol: We studied the replication of AlMV in cowpea protoplasts. However, our probes do not cross-hybridize to the mRNAs in cowpea protoplasts so we could not analyse them. In tobacco protoplasts the mRNAs are rapidly induced by salicylic acid.

Loebenstein: Is this correlated with inhibition of virus replication?

Bol: We have not studied the effect of salicylic acid on virus multiplication in tobacco protoplasts, so I do not know.

Beachy: When you looked for PR protein production at the level of messenger RNA or with antibodies in systemically infected tissues, what was the sensitivity of detection? Could you observe production if there was a hundred-fold less than in a local lesion reaction?

Bol: We analysed tobacco plants infected with AlMV which gave high yields of virus production but very mild symptoms and no evidence of PR proteins. In TMV-infected plants we see over a hundred times more PR protein mRNA than in healthy plants.

Beachy: The sensitivity of detection is important because, as the fungal and bacterial pathologists are telling us, the difference between susceptibility and resistance lies in the rate of a reaction.

Sela: In my paper (Sela et al, this volume) I also present evidence that the difference between susceptibility and resistance is in the rate of response.

Harrison: Another approach to these problems is to separate the relevant genes by conventional breeding. Perhaps Dr Gianinazzi would like to comment on that.

Gianinazzi: Among the many intraspecific and interspecific hybrids we have studied for PR-b protein production and resistance expression, we obtained a hybrid between *N. glutinosa* and *N. debneyi* that produces these proteins constitutively and shows a high intrinsic resistance to TMV. This was the only hybrid showing both characteristics that we were able to obtain in the conventional way. This hybrid is not fertile but its amphidiploid (72 chromosomes) obtained by *in vitro* culture is fertile and retains the ability to produce PR-b proteins constitutively.

We tried to cross this fertile hybrid with other tobaccos. When we produced seed that germinated—with *N. tabacum* cv Samsun (nn) or Judy's Pride Burley (nn)—we obtained a plant that was very resistant to TMV and constitutively produced PR-b proteins of both parents. Therefore there is genetic information transferred from one plant to another that makes expression of the genes producing PR-b proteins constitutive and also makes this plant resistant (Gianinazzi & Ahl 1983, Ahl et al 1983). I don't know if there is a link between these two phenomena but, from the results obtained by conventional breeding, the two properties seem to be transferred together.

Certain cultivars of tobacco, such as Xanthi nc (NN), respond to polyacrylic acid: when polyacrylic acid is injected into leaves, resistance to TMV is induced. When Xanthi nc (NN) is crossed with a variety such as Izmir (nn), which doesn't respond to polyacrylic acid, the characteristic response to this chemical is transmitted to the progeny. Genetic analysis of the gene(s) involved in this response showed that it is a dominant characteristic following a mono-genetic segregation, like the N gene, but that these two characteristics are inherited independently. When we studied the production of PR-b proteins we found that this always occurs in plants that respond to polyacrylic acid. We followed more closely the $b_1{}'$ protein which is present in Izmir (nn) and not in Xanthi nc (NN) and observed that the $b_1{}'$ gene segregated in a 3:1 ratio, independently of the other two characters. We can conclude that there is monogenetic segregation for the N gene, monogenetic segregation for the gene controlling the response to polyacrylic acid, and monogenetic segregation for the $b_1{}'$ gene. These are three different systems located on three different chromosomes which are transmitted together in all the experiments that we have done (Dumas et al 1987).

Harrison: Do you think there is a linkage between these three systems?

Gianinazzi: I don't know; there are arguments both ways. When one gene works, the other two, when present, are also expressed.

van Vloten-Doting: Do plants that are producing those PR proteins always look very sick?

Gianinazzi: I am convinced that there is antagonism between the expression of active defence mechanisms in a plant and the plant's development. Conse-quently, if we are able to introduce into plants the genetic information allowing constitutive production of PR-b proteins, and if these proteins are involved in resistance, we might well obtain a positive result from a resistance point of view, but the plant might be so unhealthy that it would be agronomically uninteresting.

Bol: It is not the PR proteins that make the plants sick. If they are sprayed with salicylic acid they are full of PR proteins but look completely normal.

Gianinazzi: In that case the plant is already well developed; when the system works constitutively plants are producing PR-b proteins when they are only two to three millimetres in size, and do so continuously as they grow. If it turns out that this system is useful in protecting plants against viruses, its expression needs to be regulated in transformed organisms to permit good plant develop-ment.

Bol: A lot of changes have occurred in your hybrids. It would be nice to have a system with only the transformed PR protein gene.

Zaitlin: What is the nature of the resistance in that hybrid? *N. glutinosa* is *NN* and the hybrid would be *Nn*. Do you get fewer local lesions on the hybrid plants?

Gianinazzi: If one infects *N. glutinosa* with TMV, local lesions of about 3 or 4 mm develop. It is very difficult to obtain an infection in the hybrid; when infection does occur, the necrotic lesions are so small that a binocular micro-scope is required to see them.

Loebenstein: Most cases in which treatment with salicylic and polyacrylic acids gives rise to activation of PR proteins and resistance involve local lesion hosts. Does this treatment not activate the host genome for resistance rather than *de novo* production of PR proteins?

Bol: Salicylic acid acts on both Samsun NN and Samsun nn. Therefore, the *N* gene is not involved in that induction. Also it acts against viruses that replicate systemically, such as alfalfa mosaic virus. Necrosis has nothing to do with that response.

Loebenstein: Do you get a reduction in virus replication in protoplasts?

Bol: The reduction is 90% in intact plants and up to 99% in protoplasts.

Loebenstein: With AlMV in tobacco the virus titre rises and then falls (Ross 1941). Is this related to PR proteins?

Bol: We don't see that phenomenon, perhaps because we are not patient enough in isolating the virus. Five days after inoculation we isolate up to 2 mg of virus per gram of leaf.

van Vloten-Doting: This fluctuation is much more pronounced in cucumber mosaic virus than in alfalfa mosaic virus. We have observed it in AlMV, however.

Zaitlin: In the AlMV it was the specific infectivity of the virus, not the amount of virus, that went down with time.

Harrison: With cucumber mosaic virus there is the complication of the satellite RNA, which can affect viral replication.

van Vloten-Doting: I don't think it is related to the satellite; even strains without satellite show such behaviour.

Zaitlin: Habili & Kaper (1981) have looked at the double-stranded RNA of the satellite and shown some correlation. It is the cyclical presence of the satellite that influences the replication of the CMV.

Loebenstein: We used quite a similar system to the one that Kaper used, with another strain of CMV, and looked for this phenomenon in the green islands. We could not find double-stranded satellite RNA in this system, so there is no evidence that the decrease in titre in the green islands in CMV-infected tobacco is related to the presence of satellites.

Harrison: We have a whole range of events in these plants. Is anybody prepared to speculate as to why they are all activated at about the same time? Is there any relevant experimental evidence?

Beachy: First we should clarify whether these responses really are induced at the same time. Dr Bol, you have the clones. Are all the mRNAs induced simultaneously?

Bol: No. The smaller mRNAs are induced a little more quickly than the larger ones. However, we are only studying the balance between degradation and synthesis. It is necessary to look at 'run-on' transcription in isolated nuclei. We need to analyse the promoter regions of all those genes to see if there are common elements responsible for the induction.

Beachy: It might be that one messenger is induced first and then its product is responsible for generating a change in calcium (or other ion) flux which induces a set of other messengers. Until the transcription assays are done very carefully under closely controlled conditions, we cannot know the answer.

Bol: With salicylic acid we induced only PR-1a and -b and PR-2, and those did not induce the other PR proteins.

Goldbach: The *Cladosporium fulvum* –tomato system could give an insight into the general mechanism. Dr De Wit (Dept of Phytopathology, Wageningen) found that if *Cladosporium fulvum* contains the avirulence gene A9 it produces a small, 28 amino acid peptide. This peptide has been fully sequenced and can be chemically synthesized. When the synthetic peptide is injected into tomato, necrosis results and phytoalexins and PR proteins are produced. Since the trigger has thus been identified, the mechanism might be solved.

van Vloten-Doting: There are still a large number of responses after the trigger.

Goldbach: Yes, but the trigger works only in the presence of avirulence gene A9 which is incompatible with resistance gene CF9, and only in this particular combination does one see necrosis. The next step is to find out what resistance gene CF9 is. Then one could try to find out what happens afterwards. The peptide produced by *Cladosporium fulvum* might be useful to those workers who aim to mimic necrosis in tomato. Addition of the peptide to protoplasts would be expected to produce a necrosis-like reaction.

Harrison: Perhaps one of the non-structural virus proteins might contain a sequence that would activate the whole series of responses.

Fritig: The 30K protein is a non-structural protein which, as we have discussed, is involved in cell-to-cell movement. Professor K. Mundry's group in Stuttgart proposes that the function of this protein could also be to overcome necrotization, which would be the general feature of most tobacco-TMV interactions. They used a *N'* gene host, *Nicotiana sylvestris*. With this host, non-necrotic wild-type TMV is known to mutate easily to local lesion variants upon *in vitro* mutagenesis. Mutagenesis trials remain unsuccessful in the opposite direction (loss of necrotization ability to yield systemic variants from local lesion types). Following an approach introduced by Kado & Knight (1966), they attempted to map on TMV RNA the ability to cause hypersensitivity on *N'* gene hosts. The data obtained (H. Nitschko et al, unpublished paper, EMBO Workshop on Molecular Plant Virology, Wageningen, The Netherlands, July 1986) suggested that uncoating the 30K cistron is necessary to permit easy mutagenesis towards local lesion variants of TMV.

Hypersensitivity is triggered by the recognition of the resistance gene product of the host by the 'avirulence' gene product of the virus. Since mutation on the 30K gene of the wild-type TMV can change a systemic infection of the host carrying the *N'* gene to a hypersensitive reaction, is the 30K gene the 'avirulence' gene of the virus? The results of H. Nitschko et al do not favour this hypothesis and suggest that the avirulence function responsible for necrotization is located somewhere else on the viral RNA and that the 30K gene of virulent virus strains has an 'anti-necrosis' activity.

Beachy: We have regenerated Xanthi nc plants with the 30K gene. If the 30K gene were the 'avirulence' gene recognized by the *N* gene, these transgenic plants would necrotize spontaneously. This is not observed; the plants look healthy. It would be interesting to repeat this experiment with the *N'N'* genotype.

Gianinazzi: S. Nicoud, a student of mine, found no difference between TMV and salicylic acid induction of PR proteins (Nicoud 1984). TMV or salicylic acid induce the same proteins in Xanthi nc (NN). I do not know what happens in Samsun.

Sela: The antiviral response is related to the *N* gene and its expression is noticeable six hours after infection by TMV in Samsun NN. There is insufficient time for the virus protein to be made to elicit the host response. So, for this response at least, something in the initial infection must act as a trigger.

van Vloten-Doting: I have done temperature shift experiments in which the coat protein gene of AlMV, at least, was translated within 10 minutes of inoculation (van Vloten-Doting 1977). In this instance six hours would be quite a long time.

Sela: If that is the case, I am mistaken. So far as I remember, we first noticed the appearance of the proteins and the RNAs in tobacco 3–6 hours after infection.

Harrison: The earlier assays were fairly insensitive, at a molecular level anyway, and more recent studies are probably more relevant.

Zaitlin: Mike Wilson has been doing experiments in which he infects the epidermis of peas, peels it off and then tags with an anti-126K antibody followed by a second, colloidal gold-tagged antibody. He finds that the 126K protein is produced within half an hour.

References

Ahl P, Gianinazzi S, Cornu A 1983 A new potential for enhancing resistance to tobacco mosaic virus in *Nicotiana* species. Neth J Plant Pathol 89:319–320

Ahl P, Gianinazzi S 1982 b-proteins as a constitutive component in highly (TMV) resistant interspecific hybrids of *N. glutinosa* × *N. debneyi*. Plant Sci Lett 26:173–181

Dumas E, Gianinazzi S, Cornu A 1987 Genetic aspects of polyacrylic acid induced resistance to tobacco mosaic virus and tobacco necrosis virus in *Nicotiana* plants. Plant Pathol, in press

Gianinazzi S, Ahl P 1983 The genetic and molecular basis of b-proteins in the genus *Nicotiana*. Neth J Plant Pathol 89:275–281

Gianinazzi S, Kassanis B 1974 Virus resistance induced in plants by polyacrylic acid. J Gen Virol 23:1–9

Habili N, Kaper JM 1981 Cucumber mosaic virus associated RNA5. VII. Double-stranded form accumulation and disease attenuation in tobacco. Virology 112:250–261

Kado CI, Knight CA 1966 Location of a local lesion gene in tobacco mosaic virus RNA. Proc Natl Acad Sci USA 55:1276–1283

Kassanis B, Gianinazzi S, White RF 1974 A possible explanation of the resistance of virus-infected tobacco plants to second infection. J Gen Virol 23:11–16

Lucas J, Henriquez AC, Lottspeich F, Henschen A, Sänger HL 1985 Amino acid sequence of the 'pathogenesis-related' leaf protein p14 from viroid-infected tomato reveals a new type of structurally unfamiliar proteins. EMBO (Eur Mol Biol Organ) J 4:2745–2749

Nicoud S 1984 Aspects moléculaires de la résistance chez les Solanacées. DEA Thesis, Dijon University, 38pp

Nyland G, Gilmer RM, Moore JD 1976 'Prunus' ringspot group. In: Virus diseases and noninfectious disorders of stone fruits in North America. Agriculture Handbook No. 437. US Department of Agriculture, p 104–132

Ross AF 1941 The concentration of alfalfa mosaic virus in tobacco plants at different periods of time after inoculation. Phytopathology 31:410–420

Sela I, Grafi G, Sher N, Edelbaum O, Yagev H, Gerassi E 1987 Resistance systems related to the N gene and their comparison with interferon. In: Plant resistance to viruses. Wiley, Chichester (Ciba Found Symp 133) p 109–119

van Vloten-Doting L 1977 Early events in the infection of tobacco with alfalfa mosaic virus. J Gen Virol 41:649–652

Weststeijn EA 1984 Evidence for a necrosis-inducing factor in tobacco mosaic virus-infected *Nicotiana tabacum* cv. Xanthi-nc grown at 22 °C but not at 32 °C. Physiol Plant Pathol 25:83–91

Mechanism of the hypersensitivity reaction of plants

B. Fritig, S. Kauffmann, B. Dumas, P. Geoffroy, M. Kopp and M. Legrand

Laboratoire de Virologie, Institut de Biologie Moléculaire des Plantes du CNRS, 12 rue du Général Zimmer, 67000 Strasbourg, France

Abstract. Active defence of plants (hypersensitive resistance) is induced by the pathogen itself. This is illustrated using the example of two almost isogenic lines of *Nicotiana tabacum* micro-inoculated with the U1 strain of tobacco mosaic virus. Necrotic stress, accompanied by metabolic alterations which are responsible for the antiviral resistance, is triggered after 33–36 hours of interaction between the plant resistance gene product and the corresponding viral avirulence gene product. The same set of metabolic changes is found in most examples of active defence during various interactions of the incompatible type. These changes are specific to the host but not to the triggering parasite. The alterations include cell wall thickening resulting from production of macromolecules, and the production of defence enzymes and proteins. Defence enzymes fall into two classes: enzymes that catalyse the production of various metabolites participating in resistance (ethylene, phytoalexins, aromatic compounds, oxidized metabolites); and direct defence enzymes (hydrolases such as chitinases and glucanases). The defence proteins include inhibitors of proteases and of polygalacturonases and pathogenesis-related (PR) proteins. Several tobacco PR proteins are in fact hydrolases. A general mechanism for the elicitation of active defence (including antiviral defence) is proposed and discussed in relation to strategies that can be used to engineer plants and confer resistance to a wide range of microorganisms.

1987 Plant resistance to viruses. Wiley, Chichester (Ciba Foundation Symposium 133) p 92–108

In this paper we shall compare aspects of the hypersensitive response of plants to viruses with features of incompatible interactions between plants and fungi. Plant resistance is usually of the passive type; most plants are non-hosts for most microbes. However, when plants are infected by parasites a variety of situations are encountered, ranging from very weak or no defence to very intense and active defence ultimately leading to resistance. The hypersensitive response is one of the most efficient natural mechanisms of defence that is induced by the infection itself. It has two main characteristics: necrosis at and around each point at which the leaf tissue was infected; and localization of the parasite to the region of each initiated infection. The cells surrounding the necrotic area undergo marked metabolic changes which are

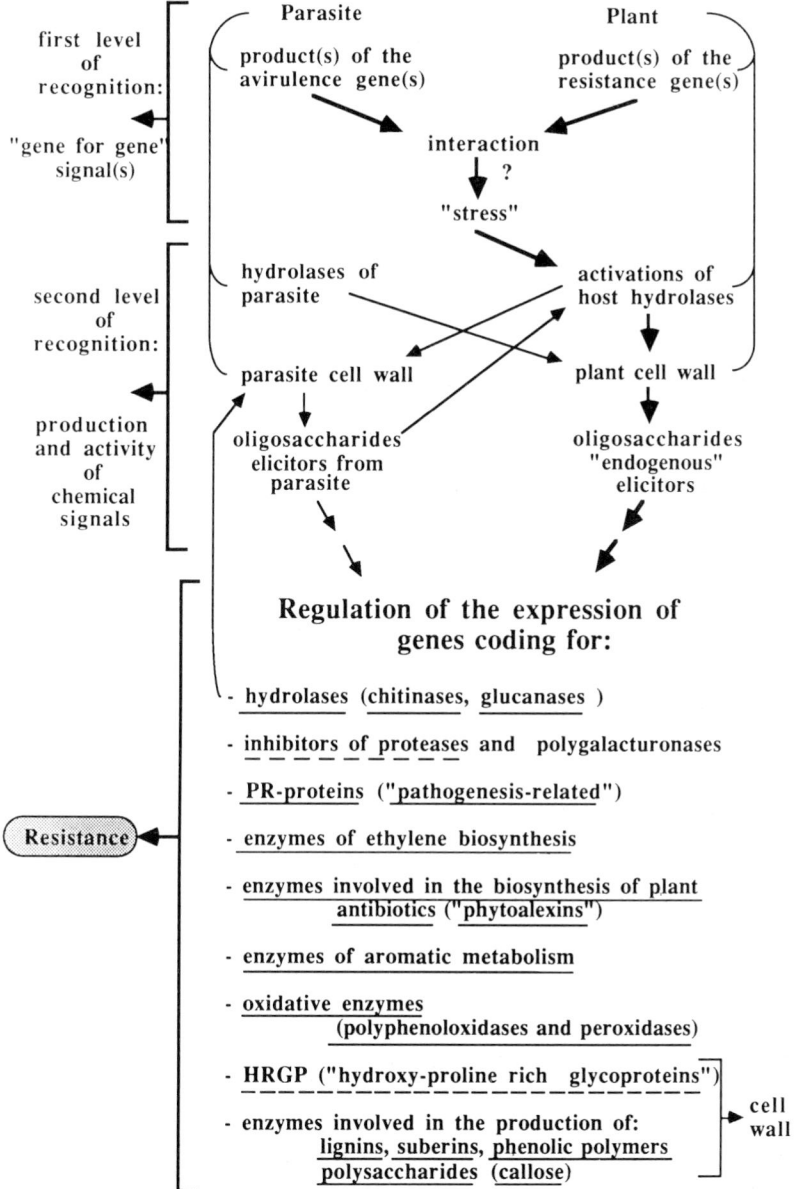

FIG. 1. Mechanism of elicitation of defence reactions in incompatible plant–pathogen interactions. Metabolic alterations that have been demonstrated in the case of hypersensitivity to viruses are underlined with continuous lines; dotted lines indicate preliminary results. The hypothetical elicitation scheme operating in virus infections is indicated with **bold face** arrows.

believed to cause, or at least to contribute to, the resistance observed. We shall show here that, even though the triggering of the hypersensitivity reaction results from a very specific gene-for-gene recognition between plant and pathogen, what the plant does thereafter to defend itself appears to be unspecific and consists of a set of metabolic alterations that are specific only to the host. This suggests that intermediate signals of host origin must be operating in all cases and that these are responsible for the induction of the metabolic changes. Hypersensitivity to plant viruses provides a simplified model system with which to search for and identify such signals. Their identification might then lead to strategies that could be used to engineer a given plant and confer resistance, not only to a given virus, but also to a wide range of pathogens.

Proposed mechanism of the hypersensitivity reaction

The proposed mechanism of the hypersensitivity reaction is outlined in Fig. 1. It consists of two successive levels of recognition signals followed by the induction of metabolic alterations responsible for resistance.

Genetic analysis of plant–pathogen interactions has shown that in most cases active defence related to incompatibility is under the control of only one host gene, the resistance gene. A parasite gene, called the 'avirulence' gene, would correspond to this resistance gene. An interaction between the products of these genes would trigger the hypersensitivity response, which is usually accompanied by a 'stress' reaction of the plant and by major metabolic alterations.

Induction of hypersensitive resistance to virus multiplication

We investigated the exact time of the appearance of resistance to virus multiplication in the hypersensitive combination *Nicotiana tabacum* cv. Samsun NN–tobacco mosaic virus (TMV). Was the resistance mechanism already operating shortly after virus inoculation, or was it induced later in the infection process? If the latter, did the onset of resistance precede or follow the stress evidenced by necrotic cell death? To answer these questions we followed virus multiplication in the hypersensitive host at intervals after inoculation and compared it with multiplication in a reference system consisting of a systemic infection, by the same TMV strain (U1), of the almost isogenic line *N. tabacum* cv. Samsun. In this type of comparison the measurements should be made on the same number of infection sites in the two hosts. However, classical mechanical inoculation creates infection sites at random on the leaves. Therefore, the exact location and number of points of virus entry can never be known for the systemically reacting host and are known only after the appearance of local lesions for the hypersensitively reacting host. To overcome this difficulty we used a micro-inoculation procedure (Konate & Fritig 1984) which yielded infection sites at predetermined locations on the leaves of the two hosts. At various intervals after the micro-inoculations the individual infection sites were assayed for virus content by an enzyme-linked immunosorbent assay (ELISA). We found large fluctuations in virus content from one infection site to another, even on the same leaf, and whatever the host or the infection period. In our comparative virus multiplication studies we analysed more than 3000 individual infection sites. This enabled us to express the results as the frequency of sites containing a given amount of virus, as will be described in detail elsewhere (G. Konate et al, unpublished).

In summary, up to 30 hours after micro-inoculation the frequency distribution curves obtained for the cultivars Samsun and Samsun NN were very similar. At about the time (33–36 hours after inoculation) that visible necrosis appeared, the rates of virus multiplication did not differ significantly between the two hosts. This observation using individual infection sites confirmed our previous results, obtained by assaying whole leaves inoculated by a classical mechanical procedure (Konate et al 1982). Once the lesions were visible on the Samsun NN leaves, the virus titres continued to increase in the infection sites of this host, but at a progressively reduced rate compared to that in Samsun infection sites. The differences between the rates of virus multiplication were demonstrated by injecting radioactive uridine around all infection sites at the end of the infection period and assaying them individually for incorporation of label into the viral RNA (Konate & Fritig 1983). The relative differences in the rates of virus multiplication during the last few hours of a two-day infection period were much greater than the relative differences between the amount of virus accumulated since the beginning of this infection period. We concluded that hypersensitive resistance to virus infection does not preexist but is induced by the infection itself. Hypersensitive resistance appears at about the same time as the necrotic symptoms and its efficiency increases with time after infection. Around the necrotic lesions there is a ring of cells about 1.0–1.5 mm in width containing detectable virus but in which virus multiplication is inhibited. With respect to the first recognition signals of Fig. 1, it appears that a period of 30–36 hours of interaction between the N gene product and the avirulence gene product of the U1 strain of TMV is required to produce the necrotic stress followed by the metabolic alterations related to resistance.

Metabolic changes characteristic of incompatible plant–pathogen interactions

How does the plant defend itself? In the lower part of Fig. 1 we list those metabolic alterations that have been most frequently reported in relation to àctive defence during incompatible plant–pathogen interactions. The best known and most studied host response is the production of phytoalexins.

These substances were discovered 45 years ago and are low molecular weight, antimicrobial compounds that plants synthesize and accumulate after exposure to microorganisms. For a long time they have been thought to be the major active principles responsible for resistance. Even though good correlations have been reported between levels of phytoalexins and the degrees of resistance, it is now generally believed that resistance ultimately results from the superposition of the various metabolic changes listed in Fig. 1. These include the production of mechanical barriers, defence proteins and defence enzymes.

Among the mechanical barriers are aromatic macromolecules such as lignins, suberins and other phenolic polymers that are produced in large amounts and incorporated into the host cell wall. Polysaccharides such as callose, a β-1,3-polyglucan, are also actively synthesized and found in several cellular compartments (cell wall, intercellular spaces, plasmodesmata). Another biochemical change contributing to mechanical barriers is the increased synthesis of hydroxyproline-rich glycoproteins which become insoluble soon after their secretion into the cell wall (McNeil et al 1984). These proteins may function in defence simply by forming a more dense, impenetrable cell wall barrier or by providing nucleation sites for the deposition of lignin, again resulting in a more protective barrier to potential pathogens. Since they possess the ability to agglutinate bacteria, these glycoproteins may also function in defence by immobilizing pathogens in the wall.

The defence proteins include pathogenesis-related (PR) proteins and protease and polygalacturonase inhibitors. Until recently, no biological function was known for PR proteins in spite of their characteristic properties and the large quantities sometimes accumulated during active defence. The PR proteins will be discussed in more detail later in this paper. Production of the inhibitors of proteases and polygalacturonases enables the plant to limit the action of enzymes that play a key role in the virulence of various pathogens.

Defence enzymes are another important category of host response. They can be subdivided into two classes: enzymes that catalyse the increased production of various metabolites participating in resistance (ethylene, phytoalexins, aromatic compounds, oxidized metabolites); and direct defence enzymes (hydrolases such as chitinases and glucanases). Chitin and glucans are major structural components of many important pests, such as insects and fungi (Bartnicki-Garcia 1968, Cabib 1987). The production by the plant of enzymes able to degrade these polymers is therefore of interest to plant pathologists.

The different metabolic changes listed in Fig. 1 are not all independent from each other; some are closely related. For instance, some of the phytoalexins are derivatives of aromatic metabolism. Also, the mechanical barrier provided by the deposition of lignin involves both the production of aromatic metabolites (substituted cinnamyl alcohols) and their polymerization by oxidative enzymes (specific peroxidases).

Metabolic changes during the hypersensitive reaction to viruses

Our group has studied some of the metabolic changes involved in the hypersensitive reaction of tobacco to TMV, namely changes in aromatic metabolism, PR proteins and, more recently, hydrolases.

We noticed a striking increase in ethanol-soluble aromatic metabolites (Fritig et al 1972) and in lignin (Massala et al 1987) which correlated well with a sharp increase in the activities of phenylalanine ammonia-lyase (PAL), cinnamic acid hydroxylase, cinnamoyl CoA ligases, ortho-diphenol-O-methyltransferases and peroxidases (Legrand et al 1976). These stimulated activities were localized in the layers of cells surrounding the necrotic lesions, and the enzymic stimulus was spreading radially in advance of the necrosis. The zones of stimulation were revealed by their fluorescence under ultra-violet light and corresponded exactly to the zones of high resistance to virus multiplication mentioned above. The difference between the enzyme levels in the fluorescent rings surrounding the necrotic lesions and the levels in healthy plants was divided by the area of fluorescence to give a value of enzymic stimulation per cell. These calculations revealed high intensities of enzymic stimulation (more than hundredfold for all the enzymes studied) which correlated well with the efficiencies of localizing different TMV strains (Legrand et al 1976).

We also studied the comparative effects on hypersensitive resistance of two competitive inhibitors of PAL, the first enzyme on the pathway. The inhibitors used were α-aminooxyacetic acid (AOA) and a structural analogue of phenylalanine, α-aminooxy-β-phenylpropionic acid (AOPP). When supplied to tobacco leaves they increased the size of the lesions for several tobacco–TMV combinations examined (Massala et al 1980, 1987), AOA being the more effective compound in weakening hypersensitive resistance. These studies demonstrated a highly specific inhibition of the phenylpropanoid pathway by AOPP. This inhibitor was also shown to be a more potent inhibitor of tobacco PAL than AOA, in vitro and in vivo (Massala et al 1987). AOA was the more efficient inhibitor of the accumulation of insoluble polymers such as lignin. The reduction of lignin synthesis therefore correlated well with the weakening of resistance. We conclude from these comparative studies that insoluble polymers such as lignin are likely to be the phenylpropanoid derivatives that participate in the mechanism of hypersensitive resistance to virus infection.

We have investigated the regulatory mechanism governing the stimulation of PAL and O-methyltransferase activities. We used density labelling to demonstrate an increased de novo synthesis of PAL and of the three O-methylating enzymes. The mechanism of synthesis of the latter enzymes was confirmed recently by a different approach in which we purified the three isoenzymes to homogeneity (Hermann et al 1987) and measured the incorporation of [^{14}C]leucine into the enzyme subunits (B. Dumas et al, unpublished results). The ultimate aim of this approach, which has now been extended to PAL and to the cinnamoyl CoA ligases, is to establish whether there is a transcriptional or translational control of the TMV-induced increase in synthesis of these aromatic metabolism enzymes. This approach has also shown that phenylpropanoid enzymes are produced in small amounts (20–50 µg per kg fresh weight of infected leaves) but exhibit very high biological activity.

Until recently this very high specific activity contrasted strikingly with the situation of the pathogenesis-related proteins. When plants are infected with viruses, viroids, fungi or bacteria, the development of symptoms is accompanied by the accumulation of soluble host-encoded proteins. Such proteins were first detected in tobacco cultivars showing hypersensitivity to TMV (van Loon & van Kammen 1970, Gianinazzi et al 1970) but have now been found in 16 plant species (van Loon 1985) in various circumstances. Since their appearance could at first only be related to pathological conditions, they were named 'pathogenesis-related' proteins. They have characteristic properties which aid in their detection: they are selectively extractable at low pH; highly resistant to proteolytic enzymes; localized predominantly in the intercellular spaces; and easily resolved by electrophoresis on polyacrylamide gels under native conditions. Despite the great wealth of information on their characteristic properties and their occurrence in large amounts in various circumstances, no biological function has yet been demonstrated for the PR proteins. In Samsun NN tobacco, ten major PR proteins could be detected and were referred to as PR proteins 1a, 1b, 1c, 2, N, O, P, Q, R and S in order of decreasing mobility on electrophoresis in native gels. The three more mobile PR proteins (1a, 1b and 1c) have been purified to homogeneity (Antoniw et al 1980) and shown to be serologically related (Antoniw & White 1983) (see also chapter by Bol et al, this volume). Since very little information was available on the physicochemical properties of the other PR proteins, we designed a preparative purification procedure using a combination of conventional and high performance liquid chromatography. The procedure, which is summarized in Fig. 2, enabled us to purify to homogeneity all known PR proteins (Jamet & Fritig 1986, S. Kauffmann et al, unpublished data). By using antibodies raised against the proteins we confirmed the serological relationships between proteins 1a, 1b and 1c, and proteins PR-2, -N and -O (Fortin et al 1985), and demonstrated new groups of serologically related proteins: PR-P and -Q, and PR-R and -S (S. Kauffmann et al, unpublished results).

Recently we looked at other metabolic changes during the Samsun NN tobacco–TMV interaction. They involve hydrolases such as chitinases and glucanases. Previous reports have mentioned highly stimulated activities of chitinase in cucumber leaves infected by tobacco necrosis virus (Métraux & Boller 1986) and of 1,3-β-glucanase in *Nicotiana glutinosa* infected by TMV (Moore & Stone 1972). We found that these enzyme activities were also

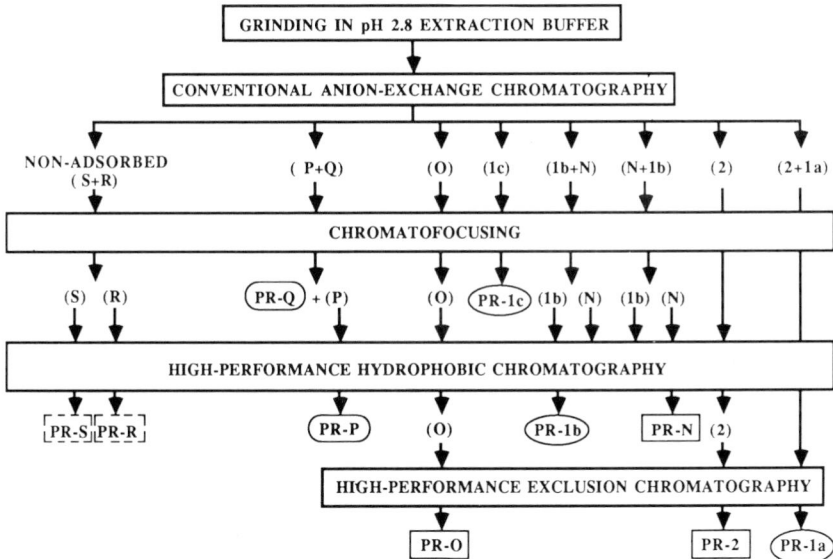

FIG. 2.   Procedure leading to the purification to homogeneity of the known tobacco
PR proteins named according to Jamet & Fritig (1986). The proteins present in the
different fractions are indicated in parentheses. Serologically related proteins are
enclosed by boxes of similar shape.

strongly enhanced in the Samsun NN–TMV combination, with time course
curves that were strikingly similar to those for the induction of several PR
proteins (Legrand et al 1987). It turned out that four chitinases — two acidic
forms with low isoelectric points and two basic ones with high pI values —
were produced upon infection. These four enzymic proteins have been purified to homogeneity and characterized. They are separable by electrophoresis
under native conditions at suitable pH and have different electrophoretic
mobilities on sodium dodecyl sulphate–polyacrylamide gels. Their relative
molecular masses (M_r) were estimated at 27 500 $\pm$ 1000 (SEM) and 28 500 $\pm$
1000 for the acidic isoforms and at 32 000 $\pm$ 1000 and 34 000 $\pm$ 1000 for the
basic isoforms of chitinase.

Several lines of evidence demonstrate that the two acidic chitinases from
tobacco are, in fact, proteins PR-P and PR-Q: (1) the acidic chitinases and the
two PR proteins displayed the same chromatographic behaviour on various
supports; (2) the estimated M_r values for the enzymic proteins purified to
homogeneity were the same as those measured for the two purified PR
proteins; (3) on native gels the purified enzymes displayed electrophoretic
mobilities characteristic of PR-P and PR-Q; (4) the antisera raised against
purified proteins PR-P and PR-Q cross-reacted with the purified enzymic
proteins; and (5) proteins PR-P and PR-Q extracted and purified by a
procedure adapted to PR proteins displayed chitinase activity. This is the first

demonstration of a biological function for proteins of the PR type. Further-more, the acidic chitinases, PR-P and -Q, are serologically closely related to the two other chitinases, which can be considered to be new basic patho-genesis-related proteins. The two basic chitinases exhibit higher M_r (32 000 and 34 000) and higher specific enzyme activity than the two acidic isoforms.

Preliminary work (Legrand et al 1987) indicated that other PR proteins of tobacco have 1,3-β-glucanase activity. The data available at present demon-strate that proteins PR-2, -N and -O are, in fact, 1,3-β-glucanases that are serologically related to a 1,3-β-glucanase of basic isoelectric point which can also be considered to be a new basic pathogenesis-related protein. In sum-mary, five of the 10 known tobacco PR proteins are polysaccharide hydrolys-ing enzymes, and this proportion is even higher if one includes the novel PR proteins with basic isoelectric points.

In hypersensitive plant–virus interactions there are additional metabolic changes that have been studied by other groups and reviewed recently (Fraser 1985, Ponz & Bruening 1986, Goodman et al 1986). Fig. 1 indicates that almost all the changes typical of incompatible plant–fungus interactions have also been found with virus infections.

Are chemical signals involved in the hypersensitive reaction to viruses?

The lack of specificity of the metabolic alterations which confer resistance contrasts with the high specificity (gene-for-gene recognition) required to initiate the hypersensitive response. This discrepancy has led many plant pathologists to search for intermediary chemical signals that would mimic incompatible plant–pathogen interactions and induce the same metabolic changes in the absence of any infection. Such regulatory molecules, called 'elicitors', were obtained first from various pathogens (for a review see Darvill & Albersheim 1984) and, later, from plant cell walls (for reviews see Darvill & Albersheim 1984, McNeil et al 1984). They are usually carbohy-drates or contain carbohydrates. Synergistic effects between elicitors of pathogen or host origin were observed in some cases. All these observations led to a hypothetical scheme (indicated in Fig. 1) involving these regulatory molecules in plant–fungi interactions (Albersheim & Darvill 1985). The stress response resulting from the gene-for-gene recognition would initiate the defence response by activating host hydrolases. Some of these hydrolases would be able to attack the host cell wall and release endogenous elicitors while others would attack the cell wall of the fungal pathogen and release exogenous elicitors which in turn could activate host hydrolases and thus amplify the host response. Furthermore, the pathogens themselves contain hydrolases that are likely to attack the host cell wall and also release endoge-nous elicitors. Therefore, it appears that a complex network of elicitors of both pathogen and host origin would act with synergistic and/or cascade effects and that this could explain the amplified efficiency of the defence responses.

Can carbohydrates interfere with a virus infection? A recent report (Modderman et al 1985) suggests that polysaccharides of plant origin may modulate hypersensitive resistance to virus infection. These authors partially digested isolated cell walls from tobacco leaves with a commercial fungal cellulase. The released carbohydrate material was fractionated by gel filtration and the resulting fractions were injected into the intercellular spaces of tobacco leaves reacting hypersensitively to TMV. A large reduction in mean lesion diameter was observed in the areas injected with carbohydrate-containing fractions.

We intend to search for putative elicitors that might be involved in hypersensitivity to viruses. Plant viruses do not contain polysaccharidic cell wall material and hydrolases, so the scheme of elicitation in Fig. 1 will need to be simplified considerably (as indicated by the bold face arrows) and include exclusively elicitors of host origin.

Conclusion

The hypersensitive reaction is switched on by the very specific recognition between the pathogen's avirulence gene product and the host's resistance gene product. Very little is known about these gene products. However, with developments in molecular biology, information is rapidly becoming available on the genome organization and expression of plant viruses and virus strains. Using site-directed mutagenesis, it should be possible to identify the avirulence gene of the virus and its product, and then the corresponding resistance gene product of the host. Even if such a resistance gene could be manipulated and introduced into normally susceptible plants, it would confer resistance against only a given virus or certain strains of this virus.

Active defence of plants is accompanied by metabolic alterations that are apparently only host specific and confer a universal type of resistance. It is noteworthy that some of the tobacco PR proteins, which were first detected in viral infections and were thought to be related to antiviral defence, are chitinases and are therefore defence enzymes directed against insects and fungi. It follows that controlling the production of regulatory molecules such as elicitors would confer resistance to various pathogens. Hypersensitivity to viruses might be a simplified model in which to identify elicitors of host origin. These elicitors could then be used to induce metabolic alterations, and the elicitors that proved most effective in inducing resistance against challenge infection by various pathogens could be selected. The gene(s) of the corresponding hydrolase(s) releasing the most efficient elicitor(s) would be useful for plant genetic engineering. If adequate regulatory sequences were used, the expression of these genes could be targeted to the sites of preferential attack by pathogens.

Acknowledgements

We should like to thank Peter Albersheim (Complex Carbohydrate Research Center, University of Georgia, Athens, USA) for very helpful discussions. We acknowledge financial support from the Centre National de la Recherche Scientifique (Grants ATP no. 5482 and no. 4483) and from the Ministère de la Recherche et de l'Enseignement Supérieur (Grants no. 83.V.0629 and no. 86.C.0952).

References

Albersheim P, Darvill AG 1985 Oligosaccharins. Sci Am 253:44–50

Antoniw JF, White RF 1983 Biochemical properties of the pathogenesis-related proteins from tobacco. Neth J Plant Pathol 89:255–264

Antoniw JF, Ritter CE, Pierpoint WS, van Loon LC 1980 Comparison of three pathogenesis-related proteins from plants of two cultivars of tobacco infected with TMV. J Gen Virol 47:79–87

Bartnicki-Garcia S 1968 Cell wall chemistry, morphogenesis and taxonomy of fungi. Annu Rev Microbiol 22:87–108

Bol JF, Hooft van Huijsduijnen RAM, Cornelissen BJC, van Kan JAL 1987 Characterization of pathogenesis-related proteins and genes. In: Plant resistance to viruses. Wiley, Chichester (Ciba Found Symp 133) p 72–91

Cabib E 1987 The synthesis and degradation of chitin. Adv Enzymol 59:59–101

Darvill AG, Albersheim P 1984 Phytoalexins and their elicitors. A defense against microbial infection in plants. Annu Rev Plant Physiol 35:243–275

Fortin MG, Parent JG, Asselin A 1985 Comparative study of two groups of b proteins (pathogenesis-related) from the intercellular fluid of *Nicotiana* leaf tissue infected by tobacco mosaic virus. Can J Bot 63:932–937

Fraser RSS 1985 Mechanisms involved in genetically controlled resistance and virulence: virus diseases. In: Fraser RSS (ed) Mechanisms of resistance to plant diseases (Adv Agric Biotechnol Ser). Martinus Nijhoff/Dr W Junk Publishers, Dordrecht, The Netherlands, p 143–196

Fritig B, Legrand M, Hirth L 1972 Changes in the metabolism of phenolic compounds during the hypersensitive reaction of tobacco to TMV. Virology 47:845–848

Gianinazzi S, Martin C, Vallee JC 1970 Hypersensibilité aux virus, température et protéines solubles chez le *Nicotiana* Xanthi nc. Apparition de nouvelles macromolécules lors de la répression de la synthèse virale. CR Hebd Séances Acad Sci Ser D Sci Nat 270:2383–2386

Goodman RN, Kiraly Z, Wood KR 1986 The biochemistry and physiology of plant disease. University of Missouri Press, Columbia, Missouri

Hermann C, Legrand M, Geoffroy P, Fritig B 1987 Enzymatic synthesis of lignin: purification to homogeneity of the three *O*-methyltransferases of tobacco and production of specific antibodies. Arch Biochem Biophys 253:367–376

Jamet E, Fritig B 1986 Purification and characterization of 8 of the pathogenesis-related proteins in tobacco leaves reacting hypersensitively to tobacco mosaic virus. Plant Mol Biol 6:69–80

Konate G, Fritig B 1983 Extension of the ELISA method to the measurement of the specific radioactivity of viruses in crude cellular extracts. J Virol Methods 6:347–356

Konate G, Fritig B 1984 An efficient microinoculation procedure to study plant virus multiplication at predetermined individual infection sites on the leaves. Phytopathol Z 109:131–138

Konate G, Kopp M, Fritig B 1982 Multiplication du virus de la mosaïque du tabac dans des hôtes à réponse systémique ou nécrotique: approche biochimique à l'étude de la résistance hypersensible aux virus. Phytopathol Z 105:214–225

Legrand M, Fritig B, Hirth L 1976 Enzymes of the phenylpropanoid pathway and the necrotic reaction of hypersensitive tobacco to tobacco mosaic virus. Phytochemistry (Oxf) 15:1353–1359

Legrand M, Kauffmann S, Geoffroy P, Fritig B 1987 Biological function of 'pathogenesis-related' proteins: four tobacco PR-proteins are chitinases. Proc Natl Acad Sci USA 84:6750–6754

McNeil M, Darvill AG, Fry SC, Albersheim P 1984 Structure and function of the primary cell walls of plants. Annu Rev Biochem 53:625–663

Massala R, Legrand M, Fritig B 1980 Effect of α-aminooxyacetate, a competitive inhibitor of phenylalanine ammonia-lyase, on the hypersensitive resistance of tobacco to tobacco mosaic virus. Physiol Plant Pathol 16:213–226

Massala R, Legrand M, Fritig B 1987 Comparative effects of two competitive inhibitors of phenylalanine ammonia-lyase on the hypersensitive resistance of tobacco to tobacco mosaic virus. Plant Physiol Biochem 25:217–225

Métraux JP, Boller T 1986 Local and systemic induction of chitinase in cucumber plants in response to viral, bacterial and fungal infections. Physiol Mol Plant Pathol 28:161–169

Modderman PW, Schot CP, Klis FM, Wieringa-Brants DH 1985 Acquired resistance in hypersensitive tobacco against tobacco mosaic virus, induced by plant cell wall components. Phytopathol Z 113:165–170

Moore AE, Stone BA 1972 Effect of infection with TMV and other viruses on the level of a β-1,3-glucan hydrolase in leaves of Nicotiana glutinosa. Virology 50:791–798

Ponz F, Bruening G 1986 Mechanisms of resistance to plant viruses. Annu Rev Phytopathol 24:355–381

van Loon LC 1985 Pathogenesis-related proteins. Plant Mol Biol 4:111–116

van Loon LC, van Kammen A 1970 Polyacrylamide disc electrophoresis of the soluble leaf proteins from Nicotiana tabacum var. 'Samsun' and 'Samsun NN'. II. Changes in protein constitution after infection with tobacco mosaic virus. Virology 40:199–211

DISCUSSION

Harrison: Dr Fritig, you observe a delay of about 30 hours after inoculation of the *N* gene tobacco plants before you begin to detect the enzymes. Is that because it takes so long for the response to start in the initially inoculated cells, or is it because many cells need to be affected to produce sufficient concentrations of the enzymes for detection?

Fritig: Professor K. Mundry (unpublished paper, EMBO Workshop on plant viruses and viroids, Strasbourg, France, July 1984) has followed the necrotization under a microscope with a camera. He found that, in *N. tabacum* Xanthi nc inoculated with TMV, the cell deaths involve a large number of cells at the same time, not just one cell after the other. Maybe a certain amount of avirulence gene product is required, but I think it is possible for the infection to move from the first infected cells to the neighbouring cells within a short time—maybe two or three hours.

Baulcombe: With several viral infections, including tobacco rattle virus, there is necrotization around the site of infection and the apparent formation of a local lesion. However, the virus is not contained within that lesion and still spreads systemically through the plant. Does this indicate that there is something beyond the so-called hypersensitive reaction?

Fritig: This is a question for general discussion. Is necrotization required for localization? One should always consider the sequence of metabolic changes: when are the metabolic alterations induced? For instance, with the same host plant (Samsun NN) and alfalfa mosaic virus, some necrotization occurs—not really local lesions, but we see phytoalexins and phenyl propanoids and yet there is no localization. I think necrotization might occur without localization because the defence is not intense enough and maybe it is not induced sufficiently in advance of infection. It is not a clear-cut situation. There is a dosage-dependent defence, which means that there is a spectrum of necrotization from weak defence to very active defence. If one were able to measure the alteration at a cellular level, one would see a very good correlation between the intensity of these alterations and the level of resistance. We have tried to do this type of measurement with the phenyl propanoid enzymes in small rings. We used a sensitive radiochemical assay which enables us to assay the enzyme in 1 mg of tissue. We calculated the average stimulation per cell and saw that in the small lesions, where there is high localization, there was greater intensity of stimulated activity of the enzymes that were assayed.

Harrison: The necrotic lesions that develop in leaves inoculated with tobacco rattle virus often develop as necrotic rings with the central part remaining green for a while after the necrosis appears. It seems that the central part, which can be quite large, is protected in some way or is not dying so rapidly.

Fritig: Is there virus inside the green part and is it in higher amounts than outside?

Harrison: Yes, there is virus there. It is hard to compare the amount from infectivity assays when the source tissue is partly dead. How can we explain the production of these necrotic rings?

Loebenstein: Perhaps one needs a build-up of the elicitor that induces the necrosis.

Harrison: Exactly. Is some kind of threshold concentration needed to induce necrosis?

Sela: There are cases where the leaves do not show necrosis but the virus is localized.

Fritig: How many cases are there like that?

Sela: There are not many, but they do exist.

Fritig: There is not just one mechanism of resistance. I have proposed a general scheme which takes into account what we know about plant–fungus interactions. I think there are other kinds of stress besides cell death. I propose that any kind of stress which occurs as a result of the plant cell–virus interaction can lead to the release or production of carbohydrates which may function as elicitors of defence. The cucumber–TMV combination is often mentioned as a rare example of virus localization without necrotization. However, in this case starch lesions are formed. As this involves alterations in the production of carbohydrates, this example may still fit into my scheme.

Sela: At what time relative to the end of virus replication do the responses that you discuss occur, Dr Fritig?

Fritig: Resistance is not a clear-cut phenomenon. The size of the local necrotic lesions grows and reaches a plateau. The time course of this lesion growth depends on the virus strain. It can be very fast but with some strains it takes longer. Therefore, some virus multiplication continues. For instance, we have seen incorporation of tritiated uridine into viral nucleic acid after seven days in the *N. tabacum* Samsun NN–TMV system.

Sela: I am interested in the other part of the growth curve, when the shift to resistance takes place. How does it correlate with the first of any of the responses that you mention?

Fritig: In a 'stimulated' cell, these alterations occur in a sequence. The ones we have studied, namely induction of enzyme activities involved in the phenyl propanoid pathway, are quite early: peroxidase, which catalyses the polymerization of the phenylpropanoid monomers to lignin, is a bit delayed.

Sela: What is early?

Fritig: When can you see necrotization? It depends how you look at it. I am not sure of the timing to the nearest hour. The PR proteins appear a little later.

Bol: Dr Fritig, do you know whether the four chitinases that you have detected have different specificities? Are some of them more important than others in the release of relevant elicitors?

Fritig: The specific activities of the basic and acidic chitinases are quite different but there are larger amounts of the acidic enzymes in the infected material, so the two acidic chitinases account for one-third of the enzyme activity.

van Vloten-Doting: Are they the same type of chitinases?

Fritig: We do not know that. We know that they are endo-enzymes, because they attack inside the polymer(s) to yield chitin fragments.

Hohn: What is the natural substrate of the virus-induced chitinases in plants? Could it be something other than chitin?

Fritig: We are not sure whether there is a substrate in the plants which could be used by the chitinases. However, chitin is a very good substrate of the induced proteins. These enzymes hydrolyse the chitin of many microbes, such as fungi which contain up to 50% chitin in their cell wall. Without exception, all insects need chitin for their development.

Hohn: Why do viruses induce the chitinases?

Fritig: Because it's a non-specific defence. There might be a substrate for these enzymes in the plants. We are looking at that.

Hohn: Why are there both acidic and alkaline chitinases?

Fritig: This is true of the glucanases as well. Maybe the plant has to set a complete pattern to satisfy all conditions from acidic to basic.

Harrison: You suggest that the basic enzyme had a basic tail. Could this kind of tail be transferred to several of these PR proteins?

Bol: For the chitinases the difference in sequence must be at the N-terminus because they have homologous C-termini. In PR-1 proteins the difference between acidic and basic proteins is at the C-terminus. There is no homology between PR-1 proteins and chitinases.

Harrison: So you don't see any general mechanism.

Bol: No, they are totally different genes.

Antoniw: Are the basic chitinases extracellular as well?

Fritig: Yes.

Antoniw: On your acidic gels, did you see any other PR-type proteins?

Fritig: Yes. There are more PR proteins which are serologically related to the acidic PR proteins. For each member of each family there is a basic counterpart.

Baulcombe: Which of these PR proteins has homology to thaumatin?

Fritig: None of them; it still has to be discovered.

Davies: Are any of the 16 PR proteins also found among the groups of proteins that appear in response to various stresses, such as heat shock, cold shock or water-logging?

Fritig: They are not the classical 'heat-shock proteins'. Some might be the same as those induced by osmotic shock: they are definitely the same as those produced during fungal infections.

Davies: Does that include the chitinases?

Fritig: Yes.

Sela: Dr I. Chet has isolated a soil fungus that biologically controls other fungal infections. He also found that chitinase and β-glucanase activities are responsible for this kind of protective mechanism.

Goldbach: In your scheme in Fig. 1, one of the most important steps is presumably the induction of the host hydrolases by polysaccharides. If this does not occur, the whole scheme fails.

Fritig: Several types of activation of the host hydrolases can be envisaged. One type might be very rapid; others require enzyme synthesis, which we have found to be rather delayed. If there is local membrane damage giving rise to electrolyte leakage there might be changes in pH very close to the cell wall. These enzymes have a pH-dependent curve with a very sharp peak and may not be active under normal conditions. If the pH is changed they could be activated in the cells that are 'stressed' and destined to necrotize. They could then attack polymeric substrates in the cells undergoing, or close to undergoing, necrosis.

This would release elicitors that would activate the defence of neighbouring cells.

Goldbach: So the activation is at the level of the protein, rather than gene expression?

Fritig: There might be a combination of the two. It might start with an activation at the protein level in the 'stressed' cells to initiate the production of elicitors, which in turn would amplify the production of host hydrolases by acting at gene expression level in neighbouring cells.

Sherwood: Has there been any speculation on the initial recognition signal between the virus and the plant? You mentioned the work of Dr P. Albersheim on the cell wall carbohydrates that can elicit phytoalexins and probably serve as the initial recognition. What about plant viruses?

Fritig: I do not think these elicitors are the avirulence gene products. There has been disagreement among people who work on elicitors. Initially the elicitors were thought to be interaction specific and race specific. If one had a given plant cultivar and an elicitor from an avirulent race of the fungus (where avirulent race means one that would lead to an incompatible interaction) then this elicitor would give the defence. I think that Dr P. Albersheim's group has shown that one can also get an active elicitor from a virulent race giving a compatible interaction with the same plant cultivar. These studies were made *in vitro* and the elicitors were obtained after drastic treatments, such as autoclaving the fungus. It is possible that this might change the conformation and the specificity of the elicitor. So far it seems that it is the release of the elicitor or the absence of release after contact of the fungus with the plant, and not the presence or absence of an active elicitor in the fungus, that is specific to the interaction.

Sherwood: What might be the elicitors for plant viruses?

Fritig: In this necrotizing system, necrotization could be the stress which activates host hydrolases. These enzymes might then attack host polysaccharides and release fragments with elicitor activity.

Gianinazzi: Is this elicitor a cell wall component?

Fritig: When I suggest that the elicitor is endogenous, I mean that it is derived from the *host* cell wall. Plant viruses have no cell wall and do not contain polysaccharides. It is true that a viral protein could be glycosylated and could then act as an elicitor, but it seems more reasonable to assume that an endogenous elicitor is involved.

Gianinazzi: If you inject cell wall extract from healthy plants into another plant, what happens?

Fritig: Modderman et al (1985) incubated a mixture of cellulase and macerozyme (commercial enzymes) with leaf tissue. The resulting fluid was injected into a local lesion host. Lesion size was considerably reduced—an indication of increased resistance of the hypersensitively reacting host. Released cell wall fragments thus affect the extent of virus infection.

Bruening: Dr Fritig, when you supplied tritiated uridine around the micro-infection site and observed how much TMV RNA was made, what happened to RNA synthesis, protein synthesis or ATP synthesis in those regions?

Fritig: There is a problem with labelling in local lesion-carrying hosts, which you mentioned in your review article (Ponz & Bruening 1986). It is difficult to label anything in a leaf carrying a local lesion, because of the high specific radioactivity of the precursor that is normally used for feeding by absorption through the petioles of detached leaves. Several authors find that most of the label is taken up in the veins and nothing is left to go to the mesophyll cells. We therefore inject the labelled compound directly into the leaf tissue. We have checked that we had comparable labelling of the ribosomal RNA in both local lesion and non-local lesion hosts. The labelling is slightly increased compared to the healthy Samsun N but not very much.

Bruening: Is this at the time when synthesis of TMV RNA is declining?

Fritig: Yes.

Bruening: Has any analysis been made of the incorporation of labelled amino acids, or of ^{32}P in general?

Fritig: We did the same kind of labelling with [^{14}C]leucine, [^{14}C]phenylalanine, and ^{14}C-labelled precursors of the phenylpropanoids, and we obtained the same results in terms of the virus.

Beachy: Therefore, whatever is happening to cause the decline of the virus is specific for the virus and not some general metabolic effect on RNA production.

Fritig: There is more metabolic activity rather than less in the stimulated cells showing resistance to virus. That has been shown by using an electron microscope to look at cells around the necrotic lesions. I don't think the reduction in TMV multiplication is because of death or decreased metabolic activity. As I have shown, PR proteins, as well as the enzymes that we have followed, are synthesized in increasing amounts.

References

Ponz F, Bruening G 1986 Mechanisms of resistance to plant viruses. Annu Rev Phytopathol 24:355–381
Modderman PW, Schot CP, Klis FM, Wieringa-Brants DH 1985 Acquired resistance in hypersensitive tobacco against tobacco mosaic virus, induced by plant cell wall components. Phytopathol Z 113:165–170

Resistance systems related to the N gene and their comparison with interferon

Ilan Sela, Gideon Grafi, Naamit Sher, Orit Edelbaum, Helena Yagev and Esther Gerassi

Virus Laboratory, The Hebrew University, Faculty of Agriculture, Rehovot 76100, Israel

Abstract. A probe made from a recombinant human β-interferon DNA detected specific bands in Southern blots of restriction enzyme-cleaved tobacco DNA. Specific tobacco DNA fragments also hybridized with a similar probe made from a clone of human $2',5'$-linked oligoadenylate (2–5A) synthetase. The expression of these plant genes was analysed by Northern blots using cDNA probes. Expression depended on the presence of the *N* gene in the tobacco cultivar. In both cases tobacco mosaic virus (TMV) infection stimulates expression. The plant β-interferon was studied further. The basal level of the relevant mRNA rapidly increases after TMV infection in *N* gene-carrying tobacco, and accumulation peaks 24 h after infection, whereas tobacco plants carrying the *n* allele are stimulated to synthesize this mRNA only about 80 h after inoculation. The plant enzyme which polymerizes ATP to antivirally active oligoadenylates was also purified from interferon-stimulated plant cells and found to resemble the human enzyme in composition and in the size of the various polypeptides, and to be serologically related to it. cDNA clones of both the plant β-interferon and the plant equivalent of 2–5A synthetase were isolated from cDNA libraries. It is concluded that the plant antiviral factor (AVF) is a type of β-interferon exerting its antiviral activity, in part, via metabolic pathways similar to those of the human interferon system.

1987 Plant resistance to viruses. Wiley, Chichester (Ciba Foundation Symposium 133) p 109–119

The activity of the plant antiviral factor, AVF, which is stimulated by virus infection, has been reviewed by Sela (1981a). In tobacco, the stimulation of AVF activity is associated with a single dominant gene, the *N* gene, which confers a type of resistance on the plant by localizing virus infection (Antignus et al 1975, 1977). AVF is not a single material but rather a family of proteins (Sela 1986), at least some of which are phosphorylated glycoproteins (Mozes et al 1978). There is evidence that a certain level of a precursor protein, pre-AVF, is present in every tobacco cultivar and that processing to activate AVF is intensive after infection, especially in *N* gene-carrying tobacco (Sela et al 1978). AVF also increases the activity of an ATP-polymerizing enzyme which produces antivirally active oligoadenylates in plants (Reichman et al 1983, Devash et al 1984, 1985).

Many parallels have been drawn between AVF and interferon (Sela 1981b). Indeed, human interferons (Orchansky et al 1982, Ogarkov et al 1984, Rosenberg et al 1985, Carter et al 1985), as well as the interferon-induced 2',5'-linked oligoadenylate (2–5A) (Devash et al 1982) inhibited the multiplication of tobacco mosaic virus (TMV), and other viruses, in plants. This paper demonstrates that TMV infection stimulates transcription of a plant gene homologous to that for human β-interferon, particularly in N gene-carrying plants, strengthening the analogy between the AVF and the interferon systems.

Detection of interferon-homologous sequences in tobacco

Northern blots of tobacco poly(A)$^+$ RNA were initially screened with probes to human interferons α and β but no homologous sequences were detected at this stage in mRNA from non-infected tobacco. However, after TMV infection, stimulation of the expression of sequences homologous to human β-interferon, 'plant β-interferon', was demonstrated.

Clearer results were obtained when we did reciprocal experiments. Plasmids carrying various inserts (α-interferon, β-interferon, γ-interferon, vH-*ras*) were Southern blotted and probed with ^{32}P-labelled first-strand cDNA made from mRNA extracted from tobacco Samsun NN 24 hours after TMV infection. Only the plasmid carrying β-interferon sequences or its excised insert bound this cDNA. Thus the stimulation of synthesis of mRNA species homologous to human β-interferon in N gene-carrying TMV-infected tobacco was corroborated. The homology is considerable, since hybridization and filter-washing conditions were relatively stringent.

The gradual appearance of the plant β-interferon mRNA species in Samsun NN after TMV infection was demonstrated by hybridizing cDNAs made from tobacco mRNA at various times after inoculation with the immobilized β-interferon DNA. Under these experimental conditions some plant β-interferon mRNA was first detected six hours after inoculation. It continued to accumulate for 24 hours, and by 48 hours after inoculation had decreased to a lower level. However, if the film was exposed for long enough, a basal level of this mRNA species could also be detected in non-infected plants. In contrast to the situation in Samsun NN tobacco, such mRNA species could not be detected in Samsun nn unless dextran sulphate was added to the hybridization medium, and even then only at 84 hours after TMV inoculation.

Subtracted cDNA enriched for TMV-stimulated species in Samsun NN accounted for 3–5% of the total cDNA population. Selected cDNA libraries were prepared from this subtracted DNA as described above and were plated out. Duplicates of the colony transfers were probed with the cDNA from which the library had been made and with the human β-interferon probe. The number of colonies hybridized to the human β-interferon probe was 25–50% of those that reacted with the cDNA probe. Hence up to 50% of the mRNA stimulated by TMV infection is β-interferon-related, which is an estimated 1.5–2.5% of the total mRNA species in Samsun NN 24 hours after inoculation.

A plant system homologous to the human 2–5A synthetase

A probe of human 2–5A synthetase hybridized to tobacco genomic DNA and to mRNA from TMV-infected *N* gene-carrying tobacco. The kinetics of expression of the gene of the plant ATP-polymerizing enzyme (APE), which is assumed to be an analogue of the human 2–5A synthetase, was studied by the reciprocal method; that is, cDNAs made to mRNA at various times after infection were used as probes. The APE gene had been expressed in *N* gene-carrying tobacco 24 hours after TMV infection. A clone carrying the APE cDNA was also isolated. This mRNA is far less abundant in tobacco than the plant β-interferon mRNA and the pertinent mRNA at 24 hours after inoculation is estimated to account for 0.03% of the mRNA population.

The ATP-polymerizing enzyme was purified from tobacco cells in suspension and was found to be interferon-stimulated. Like the human 2–5A synthetase, it consists of four polypeptides, three of which correspond in size to the human ones. At least two of the APE polypeptides reacted with antibodies to the human enzyme.

Discussion and conclusions

The activity of the plant antiviral factor (AVF) is associated with the *N* gene of tobacco and its appearance correlates with the time of establishment of virus localization (Antignus et al 1975). However, we consider the role of the *N* gene to be quantitative, determining at what time after infection, and in what quantity, AVF will be produced (or activated), rather than whether AVF is produced at all (Sela 1981a, 1987). This conclusion is based on a number of observations: (i) AVF-like substances were demonstrated in many plant–virus combinations, including *n* gene-carrying plants; (ii) the mosaic pattern of systemic TMV infection in Samsun nn consists of patches of virus-containing yellow areas and virus-free 'green islands' (Atkinson & Matthews 1967) that contain antiviral material and are resistant to superinfection; (iii) a precursor of AVF was demonstrated to occur in *Nicotiana tabacum* and *N. glutinosa*, irrespective of the presence of the *N* or the *n* allele; (iv) antimetabolites blocking transcription or translation weaken localization but never abolish it, indicating a preexisting basal level of AVF or pre-AVF.

Thus AVF resembles interferon in some respects, particularly in inducing antiviral activity in plants, although this has been disputed (Huisman et al 1985, Loesch-Fries et al 1985). The results presented here confirm that one of the plant's responses to virus infection is related to the interferon system. The characteristics of this response suggest that the 'plant β-interferon' is in fact AVF. A low basal level is maintained in non-infected plants, and this is increased by TMV infection in a manner quantitatively dependent on the *N* gene. The prompt response of the *N* gene-carrying tobacco results in localization, whereas the slow response of plants carrying the *n* allele allows virus spread but protects the plant by maintaining 'green islands'.

Not only is AVF structurally homologous to β-interferon but its mode of activity is analogous to that of the human β-interferon in at least one mechanism, namely the induction of antivirally active oligoadenylates via a homologous enzyme. The human and plant systems are however distinctive in some respects, particularly in the structure of the antiviral nucleotides. Expression of α- and γ-interferon is not stimulated by TMV infection, although some indications point to the presence of homologous sequences to those coding α-interferon in tobacco DNA and both interferons are antivirally active in plants. It should also be noted that AVF, like β-interferon (but unlike α-interferon), is a glycoprotein.

Acknowledgements

This study was supported in part by the Richard Allen Shankman Fund. We also wish to thank Dr S. Pestka for the human interferon clones and Dr J. Chebath for the human 2–5A synthetase clones.

References

Antignus Y, Sela I, Hauschner A 1975 A phosphorus containing fraction associated with antiviral activity of *Nicotiana* spp. carrying the gene for localization of TMV infection. Physiol Plant Pathol 6:159–168

Antignus Y, Sela I, Harpaz I 1977 Further studies on the biology of an antiviral factor (AVF) from virus-infected plants and its association with the *N*-gene of *Nicotiana* species. J Gen Virol 35:107–116

Atkinson PH, Matthews PEF 1967 Distribution of tobacco mosaic virus in systemically infected tobacco leaves. Virology 32:171–173

Carter WA, Swartz H, Gillespie DH 1985 Independent evolution of antiviral and growth-modulating activities of interferon. J Biol Response Modif 4:447–459

Devash Y, Biggs S, Sela I 1982 Multiplication of tobacco mosaic virus in tobacco leaf-discs is inhibited by (2′–5′) oligoadenylate. Science (Wash DC) 216:1415–1416

Devash Y, Gera A, Willis DH et al 1984 5′-Dephosphorylated 2′–5′–adenylate trimer and its analogs: inhibition of tobacco mosaic virus replication in TMV-infected leaf discs, protoplasts and intact tobacco plants. J Biol Chem 259:3482–3486

Devash Y, Reichman M, Sela I, Suhadolnik RJ 1985 Plant oligoadenylates: enzymatic synthesis, isolation and biological activity. Biochemistry 24:593–599

Huisman MJ, Broxterman HJG, Schellekens H, van Vloten-Doting L 1985 Human interferon does not protect cowpea plant cell protoplasts against infection with afalfa mosaic virus. Virology 143:622–625

Loesch-Fries LS, Halk EL, Nelson SE, Krahn KJ 1985 Human leukocyte interferon does not inhibit alfalfa mosaic virus in protoplasts or tobacco tissue. Virology 143:626–629

Mozes R, Antignus Y, Sela I, Harpaz I 1978 The chemical nature of an antiviral factor (AVF) from virus-infected plants. J Gen Virol 38:241–249

Ogarkov VI, Kaplan IB, Taliansky ME, Atabekov JG 1984 Suppression of the multiplication of potato viruses by human leukocyte interferon. Dokl Akad Nauk SSSR 276:743–745

Orchansky P, Rubinstein M, Sela I 1982 Human interferons protect plants from virus infection. Proc Natl Acad Sci USA 79:2279–2280

Reichman M, Devash Y, Suhadolnik RJ, Sela I 1983 Human leukocyte interferon and the antiviral factor (AVF) from virus-infected plants stimulate plant tissues to produce nucleotides with antiviral activity. Virology 128:240–244

Rosenberg N, Reichman M, Gera A, Sela I 1985 Antiviral activity of natural and recombinant human leukocyte interferons in tobacco protoplasts. Virology 140:173–178

Sela I 1981a Plant virus interactions related to resistance and localization of viral infections. Adv Virus Res 26:201–237

Sela I 1981b Interferon-like factor from virus-infected plants. Perspect Virol 11:129–139

Sela I 1986 Preparation and measurement of an antiviral protein found in tobacco mosaic virus. Methods Enzymol 119:734–744

Sela I 1987 Genetic traits to be manipulated in plants for virus disease resistance. In: Chet I (ed) Innovative approaches to plant disease control. Wiley, New York, p 325–335

Sela I, Hauschner A, Mozes R 1978 The mechanism of stimulation of the antiviral factor (AVF) in *Nicotiana* leaves. Virology 89:1–6

DISCUSSION

Matthews: Contrary to Dr Sela's suggestion, AVF is unlikely to have any implication for the dark green islands in a mosaic because the factor would diffuse from cell to cell, but the boundaries between dark green non-infected and fully infected cells are very sharp for several weeks.

Hohn: Dr Sela, was the AVF antiviral activity assayed on animal viruses, plant viruses or both?

Sela: None of our attempts to use AVF on an animal system worked. All three human interferons were antivirally active in plants.

Hohn: How was the antiviral activity in plants assayed?

Sela: We do a number of assays. In one we take cells into suspension, add the interferon and shake the mixture overnight. The protoplasts are prepared and then inoculated. The multiplication of TMV is measured at regular intervals. Another type of assay involves rubbing the leaf surface with interferon. We employ several kinds of interferon-containing ointments that are used to treat diseases such as herpes. After six hours the leaf is inoculated with TMV. At various intervals after inoculation discs are punched out and virus multiplication is measured by enzyme-linked immunosorbent assay (ELISA), dot-blot analysis, or infectivity.

van Vloten-Doting: Loesch-Fries et al (1985) and our group (Huisman et al 1985) have both failed to confirm protection against plant viruses by interferon.

Harrison: Did you use a similar technique to that of Dr Sela?

van Vloten-Doting: Dr Loesch-Fries followed the instructions in Dr Sela's papers very carefully. We used slightly different methods but we compared the results and there was no substantial antiviral effect of interferon. After the experiments we confirmed that the interferon was still active so there is no trivial explanation for the lack of reproducibility.

Sela: I don't know why this is. A number of other groups, including one of the leading interferon groups (Carter et al 1985) have successfully repeated my experiments. The Russian group led by Dr J.G. Atabekov have reproduced the studies with potato viruses (Ogarkov et al 1984). Golgher and Vincente in Brazil have done these experiments with other viruses.

The interferon effect is transient, which must be taken into account when using it. However, interferon does induce the 2–5A synthetase-like activity in plants, it does inhibit protein synthesis in plants, and it has a gene. Human β-interferon is the only human interferon which is a glycoprotein. I don't think it will be possible to use interferon as an antiviral agent in plants, but as far as resistance mechanisms are concerned we are close to solving this one on a molecular basis.

Sänger: Could not the debate on AVF be closed by isolating the substance and sequencing it? We had the same problem with P14, one of the PR proteins. We used 30 or 40 kilograms of infected leaves to isolate and purify the protein and to sequence it. We found that P14 is not related to any of the 4000 other proteins sequenced so far. Surely the best approach to this problem is to obtain the AVF sequence?

Sela: Yes, but that approach requires a large amount of leaves, as you say. We intend to use another method as well, namely to put this gene into a special expression vector. Perhaps we shall be able to make as much protein as we need, by that route.

Beachy: Dr Sela, you stated that the mRNA constituted half of the 5% of total RNA that was synthesized after virus infection. A number of people have used differential screening of cDNA libraries made in tobacco. They have not found any newly synthesized products except the PR proteins. We need more details—for example, the size of the RNAs, the size of the DNAs, and the stringencies of the hybridization reactions, so that we can evaluate your results.

Sela: With human β-interferon, hybridization was performed at 65°C without formamide. It can be done at 37°C with formamide.

Goldbach: Which vectors were used to clone the α- and β-interferons?

Sela: We used the SP6-derived plasmid, designed at Hoffmann-La Roche to distinguish between α- and β-interferon. The two inserts were cloned in exactly the same vector. After hybridization with the interferons the washing was of moderate stringency—65°C, 1 × SSC. For hybridization with 2–5A synthetase the stringency could be increased to 0.1 × SSC at 65°C for the DNA probe and at 85°C for the RNA probe. This suggests considerable homology, but only to a short segment of the gene; we have two different probes and we can show a better homology to one than the other.

We now have a problem of priorities. At present our efforts are directed towards sequencing AVF. We have subcloned it into M-13 and are trying to screen for a full-length isolate. This will be easy with the plant interferon but won't be simple with the plant 2–5A synthetase, where we obtained only two isolates.

Loebenstein: How do you explain that with AVF you do not obtain a dose response with respect to the inhibition of virus replication?

Sela: We do see a dose response relationship but it is unusual. This did not surprise us because the same phenomenon occurs at low dosages of interferon in animals. Down to dilutions of 10^{-6} to 10^{-8} of a standard preparation of AVF the response is normal, but then anomalies occur.

Loebenstein: On increasing the concentration of AVF, can you obtain 90% inhibition of virus replication?

Sela: The dose response is as expected when the concentration is increased. The anomaly in the dose response occurs below a certain dose level. After repeated dilution the antiviral activity is completely lost. Before this point is reached, we observe as much as 100% inhibition, which indicates only that the virus titre falls below the detection capability.

Harrison: Dr Sela, I believe that you claim to detect an inhibition of virus replication with less than one molecule of AVF per cell. What did you do to get that result?

Sela: We were purifying AVF by various steps, the final one being by HPLC. On HPLC a number of proteins were detected. We took the protein which we judged to be the most active from the HPLC pattern, ran an SDS gel, and found a single band. We sliced the gel up and showed that this band was active. We made a series of dilutions to the end-point of sensitivity. We did not have any dose response problems with the purified material. Because we could estimate from the gel how much protein was there, and we knew the end-point of activity and the molecular weight, we could estimate this end-point in terms of the number of molecules. We obtained a result which we hesitated to state— namely, less than one molecule per cell.

We assayed the diluted solution by placing it in a petri dish in a phosphate buffer, floating discs of TMV-infected leaves immediately after infection, and then measuring the multiplication of TMV over a long period. This study gives a clear dose response, with infectivity being completely prevented at a level of 0.1 molecule per cell. The problem is to calculate how many cells there are in a disc.

Harrison: What criterion of activity do you use? Does virus replication have to be reduced by 50% or 20%, say?

Sela: The criterion is the end-point of activity. In this case we didn't have a problem. We compared mock-treated leaves with leaves treated with the diluted AVF.

Harrison: So the inhibitory activity is expressed as a delay of hours before the virus content increases.

Sela: Yes. I stated previously that the plant interferon effect is very transient. Interferon in animal systems also has a transient effect. Cells don't survive the effect of interferon activity for very long. Interferon or AVF suppresses everything in the cell. After a time the animal immune system starts working. I don't know what happens in plants.

Harrison: What is the actual delay with less than one molecule of AVF per cell?

Sela: It varies from one experiment to another. With a purified AVF solution with 100 molecules or more per cell, virus is not observed at all before 96 hours. With a dilution that provides one molecule per cell, some virus is detected at 60 hours but about 50% inhibition is still observed at 96 hours.

Fraser: I have tried to calculate from your published data the relative activities of interferon and AVF in plant systems against plant viruses. The result I obtained was that interferon has a specific activity 2000 times that of AVF.

Sela: I have not done that calculation. However, I would not be surprised—interferon in plants is about 1000 times more active than it is in the corresponding animal tissues.

Harrison: Dr Loebenstein, you have worked on an antiviral substance—the inhibitor of viral replication, IVR. Are there similarities with AVF? Have you any results that might be relevant here?

Loebenstein: The work on IVR in our laboratory is done in collaboration with Dr A. Gera, Dr S. Spiegel and Dr A. Salomon. IVR is produced in protoplasts of TMV-infected Samsun NN plants. It is not released by tobacco NN protoplasts infected by cucumber mosaic virus, which does not give a localized response. IVR differs from the PR proteins and other antiviral compounds that have been discussed.

In isolating IVR, we obtained two active fractions, 26K and 57K, separated by gel filtration. We presumed that it was a dimer, and we have some evidence for this now. IVR is neither host-specific nor virus-specific. We have tested IVR on leaf discs from several hosts—tobacco, cucumber and sweet pepper tissue—infected with either cucumber mosaic virus or potato virus X.

We prepared antisera to both the 26K and 57K fractions of IVR and to the mixture. Each antiserum was active against each fraction of IVR and against the mixture. Sham-inoculation of NN protoplasts did not give any reaction with the antisera. We used the antiserum for affinity chromatography purification of

IVR and obtained a band in the 21.5K area. Thus we have a relatively clean preparation of IVR.

One difference between IVR and the PR proteins is that IVR inhibits virus replication in protoplasts, leaf discs or intact plants, even when applied up to 18 hours after virus inoculation. This makes IVR an inhibitor of replication. Another difference between IVR and most of the PR proteins is that we only detect IVR in large quantities. We have to use an incubation medium from about 1–20 million protoplasts.

Recently we found that IVR can be obtained directly from the intercellular wash fluid of infected NN leaves and also from induced resistant tissue between stripes of lesions. We took TMV-infected NN leaves five to seven days after virus inoculation and extracted the intercellular wash fluid by the techniques described by Parent & Asselin (1984). The recovered material had the same properties as IVR from protoplasts; it gave two peaks at 26K and 57K by gel filtration and cross-reacted with the antiserum prepared against the IVR from the protoplasts. The material from the intercellular wash fluid showed the same dose-responsive inhibition rate. In protoplasts we can get inhibition based on infectivity of 80%, and based on serology of over 90%.

Thus we have antisera specific to IVR and can obtain material directly from infected leaves. We are now preparing monoclonal antibodies to IVR.

Harrison: How do you get IVR into the leaf discs at different times after virus inoculation?

Loebenstein: We inoculate the leaf with virus and after several hours we punch out discs. We float these discs on either an aqueous solution or a phosphate solution containing IVR. We don't yet understand how IVR enters the leaf and how it acts against virus multiplication.

Harrison: IVR is rather a large molecule to enter the cells.

Loebenstein: It is possible that the molecule enters the intercellular space and acts there, rather than entering the cells.

Matthews: I am also concerned about the uptake of the protein by the leaf discs. When leaf discs are floated on phosphate or sulphate solutions, almost all uptake is through the cut edge. Furthermore, within a few hours of excision, cells around the border act as a metabolic trap for material which enters later. Thus even low molecular weight compounds become concentrated in a ring of cells around the edge of the disc.

Harrison: What size discs are you working with, Dr Loebenstein?

Loebenstein: We use quite large discs, one to two cm in diameter. Maybe the protein does enter through the cut edges. However, IVR is active against viruses when applied to intact plants by spraying on both sides of the leaf.

Sela: It might even be that this protein does not have to be taken up by the leaf. We know from my calculation for AVF that only one molecule per cell or even less may be required to produce an antiviral reaction. Perhaps IVR is able to trigger a minimum number of cells from the outside and to start a cascade of reactions inside the leaf.

Harrison: That might be a possibility. Dr Loebenstein, if you infect Samsun NN protoplasts with other viruses do you recover anything that resembles IVR?

Loebenstein: No, we did not try inoculation of Samsun NN protoplasts with viruses other than TMV. We did use a different system—the green island system. We took protoplasts from fully developed green islands after inoculation of cucumber mosaic virus. We recovered a compound that is similar, if not identical, to IVR.

There are correlations between NN tobacco resistance in protoplasts and IVR. Actinomycin D, applied up to 24 hours after inoculating NN protoplasts with TMV, increases the titre of TMV in these protoplasts and inhibits IVR release. Thus actinomycin D prevents virus inhibition in NN protoplasts, the virus titres (measured by infectivity or serology) increase, and the release of IVR is inhibited. There is no effect in Samsun protoplasts. Actinomycin D applied more than 48 hours after inoculation has no effect.

Sänger: What happens if the lower leaves of a plant are inoculated? Have you checked the upper leaves for the release of IVR into the intercellular space? Is there a correlation between IVR and the inhibition of virus spread?

Loebenstein: Dr P. Ahl in Switzerland used our antiserum for IVR to do this experiment. She detected IVR in the upper resistant leaves. We do not yet know if there is a correlation between the amount of IVR produced and the level of resistance. However, we do find that we can extract the largest amount of IVR from the area between the stripes. This area is also where the induced resistance is greatest: there are fewer lesions, and those lesions are smaller than in the less resistant upper leaves.

Sela: You mentioned that IVR has an antiviral effect if applied up to 18 hours after virus inoculation, whereas we find that AVF works only six hours after TMV inoculation and that the effect is greater if AVF is applied before the virus is inoculated. Do you have any other reasons for thinking that they are different, particularly as the low molecular weight (26K) IVR is the same size as the purified AVF? Do AVF and IVR cross-react serologically? Dr A. Gera asked me for a sample of AVF to test with the IVR antisera. I believe there was a weak cross-reaction.

Loebenstein: One cannot rely on molecular weight measurements because they vary according to the method of determination. I don't know if AVF and IVR are the same or different. AVF did not cross-react with our IVR antiserum. Perhaps the concentration of AVF was too low. Normally we test IVR six to seven hours after infection, but we detect antiviral activity even when IVR is applied 18 hours after virus inoculation. I suggest that AVF should be tested under conditions which preclude the possibility that it inhibits infection in the initial stages. If the antiviral activity is tested by mixing AVF with the viral inoculum or applying it shortly after inoculation, one cannot rule out inhibition of infection itself.

Sela: Dr M. Chessin from Montana collaborated with you, Dr Loebenstein. He purified the 26K IVR in the same way as I purified AVF and concluded that the 26K component of IVR is identical to AVF. I think we are dealing with two very closely related proteins which may even be identical.

References

Carter WA, Swartz H, Gillespie DH 1985 Independent evolution of antiviral and growth-modulating activities of interferon. J Biol Response Modif 4:447–459

Huisman MJ, Broxterman HJG, Schellekens H, van Vloten-Doting L 1985 Human interferon does not protect cowpea plant cell protoplasts against infection with alfalfa mosaic virus. Virology 143:622–625

Loesch-Fries LS, Halk EL, Nelson SE, Krahn KJ 1985 Human leukocyte interferon does not inhibit alfalfa mosaic virus in protoplasts or tobacco tissue. Virology 143:626–629

Ogarkov VI, Kaplan IB, Taliansky ME, Atabekov JG 1984 Suppression of the multiplication of potato viruses by human leukocyte interferon. Dokl Akad Nauk SSSR 276:743–745

Parent JG, Asselin A 1984 Detection of pathogenesis-related proteins (PR or b) and of other proteins in the intercellular fluid of hypersensitive plants infected with tobacco mosaic virus. Can J Bot 62:564–569

Analysis of the *N* gene of *Nicotiana*

David D. Dunigan, Daniel B. Golemboski and Milton Zaitlin

Department of Plant Pathology, Cornell University, Ithaca, NY 14853, USA

Abstract. The genetic history of the *N* gene, which is responsible for the hypersensitive (necrotic) response (HSR) exhibited by many species and varieties of *Nicotiana* upon infection by tobacco mosaic virus (TMV), is traced from its origin in *N. glutinosa* to its introduction into *N. tabacum* cultivars. Experiments have been designed to characterize the gene products and isolate the gene responsible for HSR. Messenger RNAs and proteins specific to HSR have been identified by utilizing the temperature-sensitive nature of the response to enrich for HSR-specific products. A minimum of four polypeptides specific to HSR were detected by *in vitro* translation of these mRNAs. Several differential hybridization protocols have been utilized in attempts to identify the gene(s) that are unique to *N. tabacum* showing HSR. Over 110 000 clones have been screened, but none were found to be specific to HSR, although several promising candidates are being examined in detail.

1987 Plant resistance to viruses. Wiley, Chichester (Ciba Foundation Symposium 133 p 120–135

One way in which plants exhibit defence against virus disease is by developing necrosis at the site of the initial infection. There is limited virus replication around these sites and the virus does not usually spread systemically; thus the plant is resistant to disease in a practical sense. Holmes (1929) described this host response with tobacco mosaic virus (TMV) infection of *Nicotiana* species, and was the first to recognize its usefulness in bioassays. This hypersensitive response (HSR) has probably been the single most studied phenomenon in plant virology, but it is still poorly understood in many respects. Many of the biochemical processes that are activated in the plant as a consequence of the virus–host interaction are known (Legrand et al 1976) but the underlying factor that *initiates* the response is not. In *Nicotiana* this trait is controlled by a single dominant gene, the *N* gene; its expression is temperature sensitive in *Nicotiana* species which carry that gene.

Strategies for the isolation of the *N* gene and the corresponding HSR-related genes must take into account the origin of these genes in the cultivars selected. This paper reviews the genealogy of the *N* gene in tobacco and outlines our strategy and progress towards the isolation of these genes.

Origin of the *N* gene in cultivars of *Nicotiana tabacum*

In Figs. 1 and 2 we outline the genetic history of the *N* gene, tracing it from its

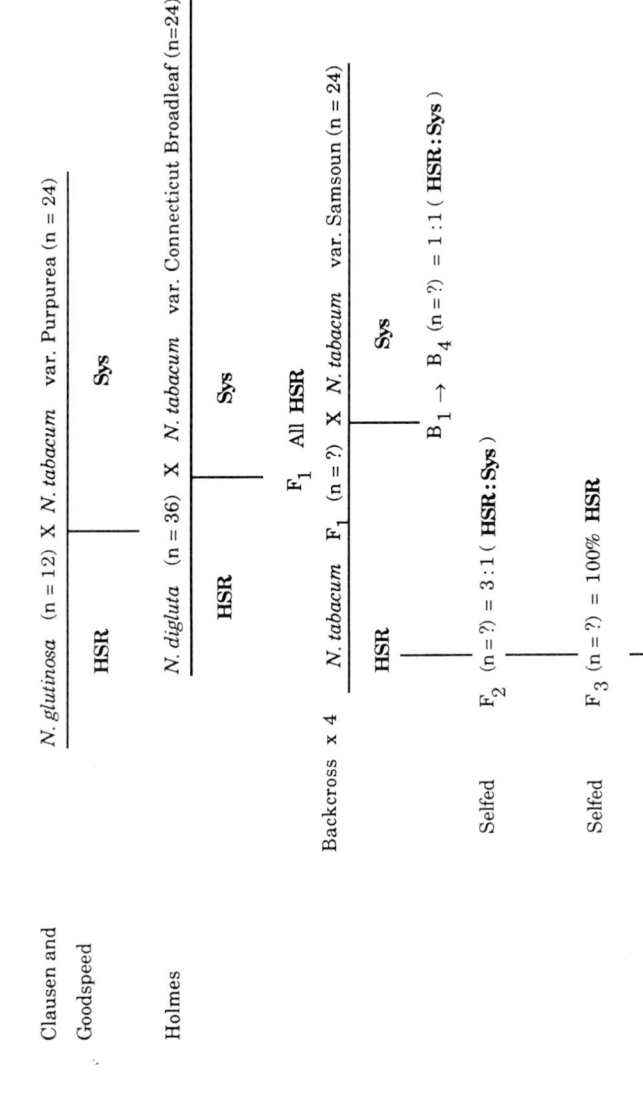

FIG. 1. The derivation of Holmes Samsoun tobacco. This figure depicts the genetic crosses used to develop the TMV-resistant variety of *Nicotiana tabacum* var. Holmes Samsoun. By convention, the female or seed parent is listed on the left and the male or pollen parent is listed on the right. n indicates the number of pairs of segregating chromosomes at meiosis; "?" indicates that the n value was not determined. HSR, hypersensitive response or necrotic response; Sys, systemic mosaic (termed a chlorotic response by Holmes). $F_{(n)}$ indicates the progeny of selfing; $B_{(n)}$ indicates the progeny of backcrosses.

1943   Gerstel

N. tabacum , F$_1$ Purpurea - Holmes Samsoun (n = 24)   X   *N. tabacum* var. Purpurea (n = 24)

HSR	**Sys**

1) *N. tabacum* , 23 bivalents
   HSR   1 monovalent, *N. tabacum*
              1 monovalent, *N. glutinosa*

2) *N. tabacum* , 24 bivalents
   HSR   1 monovalent, *N. glutinosa*

3) *N. tabacum*
   Sys

N. tabacum var. Purpurea (n = 24)   X   *N. tabacum* , F$_1$ Purpurea - Holmes Samsoun (n = 24)

Sys	**HSR**

1) *N. tabacum* , 23 bivalents
   HSR   1 monovalent, *N. tabacum*
              1 monovalent, *N. glutinosa*

2) *N. tabacum*
   Sys

FIG. 2.   Cytogenetical analysis of Holmes Samsoun-derived tobacco. The orientation of the crosses, with the female on the left and the male on the right, and the abbreviations n, HSR, Sys and F$_{(n)}$, are as indicated in Fig. 1.

origin in *N. glutinosa* to its introduction into *N. tabacum* cultivars. Clausen & Goodspeed (1925) developed the interspecific synthetic hybrid, *N. digluta*, from the cross of *N. glutinosa* ♀ × *N. tabacum* var. Purpurea ♂ which was amphidiploid. Interestingly, only three self-fertile plants were obtained and only one of these survived long enough to yield the *N* gene-containing plants used in many subsequent studies. Holmes (1938) obtained this *N. digluta* species from Clausen and Goodspeed and showed that it gave a hypersensitive response to TMV, identical to that of its *N. glutinosa* parent. He then hybridized the *N. digluta* with *N. tabacum* var. Connecticut Broadleaf. All the progeny responded to TMV infection by producing necrotic primary lesions like those of *N. digluta* and *N. glutinosa*. Backcrossing this F_1 hybrid, used as the ♀ parent, with *N. tabacum* var. Samsoun as the ♂ parent gave nearly a 1:1 (HSR:systemic) segregation pattern with respect to TMV infection response with a slight bias toward the systemic-type response. Repeated backcrossing (three additional generations) confirmed this initial segregation pattern, thus suggesting that the necrotic response had been introduced to *N. tabacum* var. Samsoun as a single dominant gene. The F_2 selfed generation of the *N. digluta* ♀ × *N. tabacum* var. Samsoun F_1 ♂ segregated 3:1 (HSR: systemic), thus fulfilling the criteria for a single dominant gene.

Gametic purity with respect to this gene was demonstrated by the more sensitive test of reciprocal backcrosses to the systemically responding Samsoun plants. The F_3 generation (apparently *NN* genotype) had segregated 339:0 (HSR:systemic) and these were reciprocally backcrossed to *N. tabacum* var. Samsoun (genotype *nn*). Without exception, these progeny were of the HSR type. Holmes had thus established a homozygous *NN* genotype variety of Samsoun tobacco, which is now referred to as Holmes Samsoun or Holmes Samsun (Gerstel 1943). (Although we cannot document precisely when the change occurred, the spelling has apparently been changed to Samsun, a name now used for these varieties; Wolf 1962.) It is important to point out that other horticultural varieties of *N. tabacum* (Connecticut Broadleaf and White Burley) did not give such clear-cut segregation patterns when similar crosses with F_1 were made (Holmes 1938). This suggests to us a possible incompatibility of the *N. glutinosa* chromosomes with the chromosomes of these varieties of *N. tabacum*. On the other hand, from Holmes' data, it would seem that there is some compatibility between the chromosomes of *N. glutinosa* and *N. tabacum* var. Samsoun, even though they are not considered to be close relatives (Valleau 1952).

Mallah (1943) demonstrated non-pairing of certain chromosomes of Holmes Samsoun during meiosis (non-conjunctional pairing) and suggested that the *N. glutinosa* chromosome containing the *N* gene had completely displaced an *N. tabacum* chromosome. Gerstel (1943) confirmed Mallah's hypothesis correlating the non-conjunctional pairing of *N. glutinosa*-derived plants (*N. digluta* and Holmes Samsoun) to the necrotic response to TMV infection. The HSR-type plants had been derived from the cross of F_1

Purpurea–Holmes Samsoun ♀ × Purpurea ♂, where Purpurea is the 'standard' *N. tabacum* variety (Mallah 1943). The HSR-type plants were of two chromosomal patterns: (1) 23 bivalents plus two monovalents, the two monovalents being derived from *N. tabacum* (T) and *N. glutinosa* (G); and (2) a trisomic with 24 bivalents plus one unpaired monovalent (G). When the reciprocal backcross was analysed (Purpurea ♀ × F$_1$ Purpurea–Holmes Samsoun ♂), only 23 bivalents plus two monovalents (T and G) were observed. The remarkable conclusion of these experiments is that the interspecific synthetic hybrids containing the *N* gene have been obtained by stable chromosome substitution, and that the *N. glutinosa* chromosome, although structurally unfit for chromosome pairing, is physiologically compatible when in *N. tabacum*. These derived plants are morphologically identical to the *N. tabacum* parents. The question arises of why, with repeated backcrossing and selfing of the *N* gene-derived *N. tabacum*, this entire chromosome has been retained. This observation seems to explain the ease of the *N* gene 'introgression' reported by Holmes (1938). Gerstel (1948) later observed that an exchange of segments between the H chromosomes of *N. tabacum* and *N. glutinosa* can occur.

Gerstel (1945) established by monosomic analysis that the chromosome substitution, observed by non-conjunctional pairing of Holmes Samsoun tobacco, was for the H chromosome. This is remarkable because the monosomic plant with the Hg chromosome (H chromosome derived from *N. glutinosa*) is viable. This proves that the Hg chromosome contains the genetic material normally provided by the H chromosome. A functional role for the Hg chromosome is supported by the fact that microspores with 23 chromosomes are rarely functional (Clausen & Cameron 1944), though some nullisomic plants of *N. tabacum* have been reported (Gerstel & Parry 1973, Mattingly & Collins 1974). This functional homology is important because *N. glutinosa* and *N. tabacum* are not thought to be close relatives.

In the experimental system that we have chosen we are comparing the reaction of two varieties of *N. tabacum* to TMV infection. One variety (Xanthi nc) gives HSR to severe strains of the virus (we use the U1 strain) and thus is the *NN* genotype. The other variety is Turkish Samsun and reacts by producing a systemic chlorotic mosaic (Sys); this plant is the *nn* genotype. It is our goal to isolate the *N* gene and other genes associated with the HSR (Legrand et al 1976). Obviously, any strategy for isolating the gene will depend on assumptions about the relationship of the *N* gene to possible corresponding alleles of the wild-type (Turkish Samsun) plant. We have assumed that the *n* allele is absent (null) in the Turkish Samsun variety. Given the genetic evidence, we recognize this may not be so, and thus we might have to change our strategy for *N* gene isolation.

Are there allelomorphs to the *N* gene?

In some species of *Nicotiana*, certain strains of TMV induce a systemic disease, whereas others induce HSR. This plant response is considered to be under the control of a putative allele of the *N* gene, the *N'* gene. Valleau (1952) has postulated that the *N* allele is dominant over the *N'* and *n* alleles; the inheritance pattern suggests these alleles are located on homologous chromosomes (Valleau 1943), and that they are allelomorphs. *N. tabacum* is an amphidiploid whose progenitors were *N. sylvestris* and *N. tomentosiformis* (Gerstel 1960, Sheen 1972, Gray et al 1974, Okamuro & Goldberg 1985). Weber (1951) has established that *N. sylvestris* exhibits HSR with certain mild strains of TMV, but responds systemically with some severe strains (such as the U1 strain), and this response type is inherited via the *N'* allele. *N. tomentosiformis* reacts systemically to both mild and severe strains of TMV and has the *n* allele. The *N* allele is dominant over both the *N'* and *n* alleles, and the *N'* allele is dominant over the *n* allele. Many examples of both *N'*- and *n*-type varieties of *N. tabacum* are known (Weber 1951); the only *N*-type varieties known are interspecific synthetic hybrids, as discussed above.

Clearly, it is not known precisely how these alleles relate to each other in *Nicotiana*. Melchers et al (1966) have found a spontaneous mutant of *N. tabacum* var. Samsun (genotype = *nn*) named Samsun EN, which gives a necrotic reaction to mild strains of TMV, as do the *N'* varieties. Since *N. sylvestris* (genotype = *N'N'*) is a progenitor of *N. tabacum*, the Samsun EN may have arisen as a back-mutation to its original state; that is, *n* to *N'*. If this were true, *N'* and *n* would be established as allelomorphs. The *N* allele of *N. tabacum* var. Xanthi nc was shown to segregate independently of the *N'* alleles of either *N. tabacum* var. Java or *N. tabacum* var. Samsun EN. However, this does not exclude allelomorphism of the *N* and *N'* genes because it is not known what the nature of the relationship of the *N* gene to the H chromosome is in these plants; that is, is the *N* gene introgressed into the *N. tabacum* H chromosome or is it a stably maintained *N. glutinosa* Hg chromosome?

These data suggest, but do not prove, that *N*, *N'* and *n* are allelomorphs. Indeed, Fraser (1985) proposes that when a dominant resistance gene is derived from another species, it may be expected that the wild-type allele will be null, in this case the *N'* and *n* alleles. Yet, can both *N'* and *n* be alleles and be null? The question then arises as to the molecular meaning of a recessive allele. Recessiveness may result from any one of a broad spectrum of gene expression dysfunctions, besides being null, where there is no corresponding DNA sequence in the wild-type genome.

Isolation of HSR-specific messenger RNAs and proteins

Our first approach to the study of the *N* gene was to examine the specific mRNAs and *in vitro* translation-derived proteins directly related to HSR. The results of these studies (Smart et al 1987) will only be outlined here. We took advantage of the temperature sensitivity of the response to stimulate the synthesis of mRNAs directly related to the HSR, and to isolate and translate them *in vitro*. We compared the proteins synthesized with those produced by a series of non-HSR-induced control mRNAs. TMV-infected *N. tabacum* var. Xanthi nc plants were grown at 31 °C after inoculation and were shifted to 25 °C for eight hours, after which RNA was extracted for use as an mRNA in *in vitro* translation. This time was selected because it is before the onset of necrosis, which normally starts at about eight to nine hours under these conditions. RNA was also extracted and translated from a series of control plants selected to allow for mRNAs and proteins stimulated as a result of either virus infection, temperature shift or plant genotype. At least four polypeptide translation products specific to the HSR plants were observed. Their approximate M_r values were 60 000, 49 000, 35 000 and 33 000 (p60, p49, p35 and p33). It should be stressed that these four proteins represent the *minimum* number of HSR-affected proteins because of the one-dimensional gel system used for analysis. Very minor products, and/or those which would migrate with normal host proteins on the gel, would not be detected. We do not know the nature of these protein products, but it is not inconceivable that p35 and p33 might be pathogenesis-related proteins (van Loon 1985). One of the HSR-enhanced polypeptides might be the β-gluconase observed in HSR tissues (Mohnen et al 1985).

Attempts to isolate HSR-specific gene(s)

All our efforts so far have been focused on the use of a differential hybridization method to identify the gene(s) selectively expressed in *N. tabacum* after the induction of HSR by TMV. Two *N. tabacum* var. Xanthi nc-specific libraries were constructed. A total genomic DNA library was constructed in a Charon vector (Loenen & Blattner 1983) and a cDNA library was made by G-C tailing cDNA synthesized from Xanthi nc poly $(A)^+$ RNA and cloned into the *Pst*I site of pBR322 (Maniatis et al 1982). Differentially expressed HSR-specific RNAs are defined as those present in TMV-infected Xanthi nc temperature-shifted plants but not in TMV-infected temperature-shifted Turkish Samsun. LiCl-precipitable cytoplasmic RNA was isolated from each of these plants. Polyadenylated RNA was purified by one cycle of oligo(dT)-cellulose chromatography and the integrity of the purified poly(A)$^+$ RNA was examined by *in vitro* translation. Both libraries were screened by hybridization of radiolabelled probes derived from the poly(A)$^+$ RNAs to filter-bound plasmid or phage DNA. Approximately 75 000 clones from the genomic DNA library were screened. Hybridization of each probe to replica filters allowed for the selection of 63 clones that hybridized differentially. These clones were amplified and retested by DNA dot-blot analysis. Unfortunately, none of these clones hybridized differentially on the dot blot. Subsequently, the cDNA library was screened in the same fashion. Roughly 17 000 colonies were screened with the resultant isolation of 20 differentially expressed clones. Again, dot blots of the isolated clones failed to reveal any clones that were specific to HSR.

It is obvious that the RNA populations being compared here are very similar and that a large number of clones must be screened to detect differentially expressed messages. The mRNAs that are expressed only in the HSR plants might be too rare to be detected using this approach. By enriching the mRNA population for those genes that are selectively expressed, it should be possible to decrease the chances of incorrectly selecting genes common to both types of plants. This was done by exhaustively hybridizing [^{32}P]cDNA prepared from the HSR plant with the poly(A)$^+$ RNA from the non-responding plant and eliminating the hybrids, and isolating the remaining single-stranded cDNA from hydroxyapatite columns. A hundredfold excess of poly(A)$^+$ RNA to cDNA was used and the hybridization reactions were carried out to Rot = 1000 (Britten et al 1974). About 15% of the cDNA remained single-stranded. This cDNA was used as a probe. Approximately 25 000 clones from the genomic library were screened. As before, none was found to be specific to HSR. When the cDNA library was screened with the enriched cDNA probe, numerous positive clones were detected. These were compared with replica filters that were probed with non-enriched cDNAs made from the two different poly(A)$^+$ RNA populations. There are 11 clones, out of approximately 18 000 screened, that hybridize both with the hydroxyapatite-selected cDNA and preferentially with cDNAs from the HSR plant. These are currently being re-examined. However, the limited success so far in these studies suggests that alternative strategies for the isolation of HSR-related genes may have to be developed.

Acknowledgements

The studies described were supported in part by grant 84–09851 from the National Science Foundation and a grant from the Cornell Biotechnology Program which is sponsored by the New York State Science and Technology Foundation and a consortium of industries. David Dunigan holds a fellowship from the Cornell Biotechnology Program.

References

Britten RJ, Graham DE, Neufeld BR 1974 Analysis of repeating DNA sequences by reassociation. In: Colowick SP, Kaplan NO (eds) Methods in enzymology. Academic Press, New York, vol 29: 363–418
Clausen RE, Cameron DR 1944 Inheritance in *Nicotiana tabacum*. XVIII Monosomic analysis. Genetics 29:447–477
Clausen RE, Goodspeed TH 1925 Interspecific hybridization in *Nicotiana*. II. A tetraploid *glutinosa–tabacum* hybrid, an experimental verification of Winge's hypothesis. Genetics 10:278–284

Fraser RSS 1985 Genetics of host resistance to viruses and of virulence. In: Fraser RSS (eds) Mechanisms of resistance to plant diseases. Martinus Nijhoff/Dr W Junk Publishers, Dordrecht, p 62–79

Gerstel DU 1943 Inheritance in *Nicotiana tabacum*. XVII. Cytogenetical analysis of *glutinosa*-type resistance to mosaic disease. Genetics 28:533–536

Gerstel DU 1945 Inheritance in *Nicotiana tabacum*. XIX. Identification of the *tabacum* chromosome replaced by one from *N. glutinosa* in mosaic-resistant Holmes Samsoun tobacco. Genetics 30:448–454

Gerstel DU 1948 Transfer of the mosaic-resistance factor between H chromosomes of *Nicotiana glutinosa* and *N. tabacum*. J Agric Res 76:219–223

Gerstel DU 1960 Segregation in new allopolyploids of *Nicotiana*. I. Comparison of 6× (*N. tabacum* × *tomentosiformis*) and 6× (*N. tabacum* × *otophora*). Genetics 45:1723–1734

Gerstel DU, Parry DC 1973 Production and behavior of nullisomic S in *Nicotiana tabacum*. Tob Sci 17:78–79

Gray JC, Kung SD, Wildman SG, Sheen SJ 1974 Origin of *Nicotiana tabacum* L. detected by polypeptide composition of Fraction I protein. Nature (Lond) 252:226–227

Holmes FO 1929 Local lesions in tobacco mosaic. Bot Gaz 87:39–55

Holmes FO 1938 Inheritance of resistance to tobacco-mosaic disease in tobacco. Phytopathol 28:553–561

Legrand M, Fritig B, Hirth L 1976 Enzymes of the phenyl-propanoid pathway and the necrotic reaction of hypersensitive tobacco to tobacco mosaic virus. Phytochemistry (Oxf) 15:1353–1359

Loenen WAM, Blattner FR 1983 Lambda Charon vectors (Ch 32, 33, 34 and 35) adapted for DNA cloning in recombination-deficient hosts. Gene (Amst) 26:171–179

Mallah GS 1943 Inheritance in *Nicotiana tabacum*. XVI. Structural differences among the chromosomes of a selected group of varieties. Genetics 28:525–532

Maniatis T, Fritsch EF, Sambrook J 1982 Molecular cloning: a laboratory manual. Cold Spring Harbor Laboratory, Cold Spring Harbor, New York, p 230–242

Mattingly CF, Collins GB 1974 The use of anther-derived haploids in *Nicotiana*. III. Isolation of nullisomics from monosomic lines. Chromosoma 46:29–36

Melchers G, Jockusch H, Sengbusch PV 1966 A tobacco mutant with a dominant allele for hypersensitivity against some TMV-strains. Phytopathol Z 55:86–88

Mohnen D, Shinshi H, Felix G, Meins, Jr F 1985 Hormonal regulation of β 1, 3-gluconase messenger RNAs in cultured tobacco tissues. EMBO J 4:1631–1635

Okamuro JK, Goldberg RB 1985 Tobacco single-copy DNA is highly homologous to sequences present in the genomes of its diploid progenitors. Mol Gen Genet 198:290–298

Sheen SJ 1972 Isozymic evidence bearing on the origin of *Nicotiana tabacum* L. Evolution 26:143–154

Smart TE, Dunigan DD, Zaitlin M 1987 *In vitro* translation products of mRNAs derived from TMV-infected tobacco exhibiting a hypersensitive response. Virology 158:461–464

Valleau WD 1943 The relative positions of the N and N′ factors on *Nicotiana tabacum* chromosomes. Phytopathol 33:14

Valleau WD 1952 The evolution of susceptibility to tobacco mosaic in *Nicotiana* and the origin of the tobacco mosaic virus. Phytopathol 42:40–42

van Loon LC 1985 Pathogenesis-related proteins. Plant Mol Biol 4:111–116

Weber PVV 1951 Inheritance of a necrotic-lesion reaction to a mild strain of tobacco mosaic virus. Phytopathol 41:593–609

Wolf FA 1962 Classification and grading of aromatic tobaccos (Chap 6). In: Aromatic or oriental tobaccos. Duke University Press, Durham, p 90–103

DISCUSSION

van Vloten-Doting: Dr Zaitlin, you searched for the *N* gene messenger RNA and protein by looking at the temperature shift. I do not understand the rationale behind this. If the protein is temperature sensitive, both the RNA and the protein would probably be made at whatever temperature you chose, but would not be functional. Since you used SDS gels to detect the proteins, why did you expect to see a difference in protein levels?

Zaitlin: We were looking for enhancement of messengers which would be reflected in the protein products by *in vitro* translation. We did see some *N* gene-related products (Smart et al 1987). As you say, there is a possibility that the proteins themselves could be temperature sensitive, and are being produced all the time.

van Kammen: How specific is the reaction of the *N* gene to virus infection?

Zaitlin: It is possible that there are a number of triggering mechanisms to a necrotic reaction. We have discussed the *N* gene plant. Other plants, even nn tobacco, can be induced to form necrosis, perhaps by a somewhat different initial mechanism, but probably ending up with the same series of reactions which produce necrosis.

Loebenstein: Is the *N* gene specific for TMV? For example, tobacco necrosis virus gives a localized reaction in Samsun. Does the *NN* genotype give a localized reaction, rather than a necrotic reaction, against a virus other than TMV? I do not think it does.

Zaitlin: TMV interacts with its host in some way and only this host genotype will produce necrosis. That does not preclude other types of interactions in other genotypes with other viruses that can also produce necrosis.

Loebenstein: The question was whether the *NN* genotype is specific to TMV. I think the answer is yes. There are other genes in other hosts which give a localized necrotic reaction against other viruses. I have not come across evidence that a certain gene which localizes one virus in host A gives the same reaction against virus B.

Zaitlin: That simply means that virus B does not have the capacity to interact with the resistance gene in host A. In the case of the *N'* gene, characterized in *N. sylvestris*, some strains of TMV can interact to form necrosis, whereas other strains cannot. The important point is that there is a host–virus pair involved in the interaction which leads to localization.

Harrison: Dr Zaitlin, your study is not of the primary triggering event. Rather, you are looking at the later consequences of this interaction—such as the induction of the proteins that Dr Fritig described in his paper (Fritig et al, this volume). You are trying to obtain clones that specify the mRNAs involved. Do you already have some of these clones?

Zaitlin: We have only antiserum to the PR-1 group, the low molecular weight class of PR proteins. Some of our proteins are PR proteins.

Harrison: You have described one approach to finding viral resistance genes. The work on differences in the components of extracts of resistant and susceptible plants described by Bruening et al (this volume) might lead to another kind of study.

Bruening: Dr Zaitlin has raised the problem of not having isogenic pairs of plants. This means that this sort of approach is difficult.

Zaitlin: The ideal method would be to have a gene interruption strategy, putting a transpositional element in the *N* gene when you have it in the heterozygous state.

Harrison: In how many kinds of plants are isogenic lines available?

Fraser: We have so-called isogenic lines in tobacco, tomato and pepper, for example. However, even after eight or ten backcrosses, there is still between 0.1 and 0.4% of non-isogenic material present. Therefore, the total possible number of *gene* differences is quite large. We should use the term 'nearly isogenic'.

Beachy: For Samsun vs Samsun NN there was a chromosome transfer. Therefore the plants are absolutely not isogenic. Is there another plant pair which is more nearly isogenic? Are Xanthi and Xanthi NN from the same origin? What about Xanthi nc?

Nishiguchi: Xanthi NN is produced by the cross between Xanthi and Vamorr 48. Vamorr 48 has the *N* gene from *N. glutinosa.* Thus the *N* gene of Xanthi NN is derived from *N. glutinosa* via Vamorr 48 (Kubo et al 1971).

Zaitlin: That probably occurs by the same route as I described.

Beachy: What is the origin of the *N'* gene?

Gianinazzi: It comes from *N. sylvestris.*

Beachy: Could the *N'* and its parent be isogenic?

van Vloten-Doting: We have a problem in our definitions of isogenic. One definition is that the isogenic pair is identical except for one gene. However, in this context, isogenic has been used to describe plants that can be calculated to contain more than one difference.

Fraser: That is correct: the plants are not isogenic after four to eight backcrosses and could differ at several hundred loci.

Beachy: Dr Zaitlin has drawn our attention to literature demonstrating that the *N* gene is the result of a chromosomal transfer. Therefore we do not have an isogenic pair of plants. However, I wonder about the *N'* plant.

Gianinazzi: In what is usually called *N. tabacum* variety Xanthi nc (NN), there is an *N* gene which comes from the H chromosome of *N. glutinosa.* Transfer of the H chromosome of *N. glutinosa* into Xanthi nc (NN) was demonstrated by the early histological study by Gerstel (1945). *N. tabacum* var.

Xanthi nc (NN) comes from the cross between *N. sylvestris* and *N. tomentosiformis* and a subsequent cross with *N. glutinosa*. Therefore, Xanthi nc (NN) could have *N* and *N'* genes, but as there is a lack of information, the exact history is not well known.

Zaitlin: A number of the papers reporting these genetic details did not study the cytogenetics of the phenomenon.

Harrison: Bearing in mind that the hypersensitive response involves effects on the cell wall (see Fritig et al, this volume), has anybody compared the cell wall composition of plants with and without the *N* gene?

Fritig: We are hoping to do that study in collaboration with Professor J.P. Joseleau in Grenoble.

Beachy: Perhaps it would be useful to stop working on these special genotypes of *Nicotiana* and study *Chenopodium*, in which there are necrotic reactions with many different viruses.

van Vloten-Doting: Even in *Chenopodium* it would not be correct to say that all the viruses interact with the plant by the same mechanism. There may be a lot of different genes, each reacting with a different virus. From one virus, which produces a systemic chlorotic infection, it is possible to make a variety of mutants that induce necrotic local lesions. I think it is an oversimplification to suggest that these are all due to the same interaction with the host.

Beachy: There is no evidence for either situation. The problem is that we have a bad genetic system. Also we know nothing about the induction of these responses. We cannot say whether there is only one type of interaction or not. In looking for a new system to study, we should consider differences in the hosts as well as in the viruses.

Harrison: It appears that the specificity of the hypersensitive response lies in the nature of the initial recognition between the virus or viral products and the plant. The series of events that this interaction initiates may be similar for a wide range of pathogens. Evidently it is not too difficult to characterize these later events, but we are a long way from understanding the initial interaction.

Zaitlin: The interaction might not even be with a gene product. It could be an interaction with a region of the host genome which interrupts a particular function.

Harrison: Dr Goldbach referred earlier to a small peptide involved in disease caused by *Cladosporium fulvum* (p 89). Many small peptides could be produced from the viral genome, so these might be candidates for trigger substances.

van Vloten-Doting: These peptides could be degradation products from the coat protein. Is that peptide from *Cladosporium* a primary product, Dr Goldbach?

Goldbach: That is not known. I would expect it to be a proteolytic processing product because the corresponding gene would be very small. Dr De Wit has produced antibodies. Perhaps the peptide antibodies will cross-react with proteolytic breakdown products of some viral protein. It is informative to compare the data obtained in this fungal system with that from virus systems.

Harrison: In *Cladosporium*, is this peptide the only one which initiates the hypersensitive response, or might there be others?

Goldbach: The peptide was purified by HPLC and detected by the observation of this necrotic action. Only one peak with necrotic activity was found and it appeared to be the peptide. There is no evidence for any other trigger. This peptide is produced in compatible situations as well. It is unusual in that it is a normal product of the pathogen, yet it should be recognized by the plant to result in incompatibility.

Sela: Dr Zaitlin's comment that the plant–virus interaction might not be with a plant gene product is important. It is possible that the *N* gene is not a structural gene but is just a regulatory sequence of nucleotides that has been affected by viral protein or RNA. I am not convinced that the *N* gene is a classical gene, because the only evidence for it is the possibility of creating phenotypes after classical crosses.

van Vloten-Doting: Since the hypersensitive response is temperature sensitive at only 32 °C, it seems likely that a plant protein is affected. When the temperature transition is low for the temperature-sensitive mutants of bacteriophages, a protein is normally involved. When an RNA is involved the transition temperature is often higher and it is more difficult to demonstrate the temperature-sensitive character of the response.

Zaitlin: That is a good point.

Harrison: It wouldn't rule out the involvement of a small peptide, would it?

van Vloten-Doting: No, it suggests that it's something that has a well-defined tertiary structure which can be distorted by raising the temperature by only a few degrees.

Baulcombe: But the temperature effect could be on a regulatory protein which is affecting *N* gene expression. Therefore there could be a regulatory protein affecting the *N* gene expression which produces this hypothetical regulatory RNA. This would still account for the temperature sensitivity.

van Vloten-Doting: That would be possible—a type of cascade response.

Zaitlin: On the other hand, the virus has to interact with the host genome. Perhaps the *N* gene affects that interaction. There is information about the *N'* gene: the sequences that interact to induce the necrosis of *N. sylvestris* are known to be in the 3′one-third of the genome.

Sela: *N'* is just a name. Are *N* and *N'* related in any way?

Gianinazzi: *N'* is more than just a name because it doesn't have the same effect as the *N* gene on the different strains of TMV (Boudon et al 1970). It comes from *N. sylvestris*. *N'* is allelic with the *N* gene. There is also another '*N*' gene, the '*N*' gene of *N. rustica* which is different from the *N'* and *N* genes. It must have a different origin (see Dumas & Gianinazzi 1986).

Zaitlin: Melchers et al (1966) obtained Samsun EN, which was a reversion to a type which gives a hypersensitive response to mild strains, similar to the response of *N. sylvestris*.

Bol: Is there extensive homology between the *glutinosa* genome and the Samsun genome? Could you use a *glutinosa* probe to identify *glutinosa*-specific mRNAs in Samsun NN?

Zaitlin: It surprises me that we have not found something like that already. If we had a complete *N. glutinosa* chromosome, why did we not find some unique *glutinosa* genes in our search?

Bol: Have you made cDNA probes to poly(A)$^+$ RNA from *N. glutinosa* and used them to screen your Samsun NN library?

Zaitlin: We haven't done that.

Sela: I would not be surprised if mRNA from TMV-infected *N. glutinosa* will detect homologous sequences in *N. tabacum*, because those genes on the *glutinosa* chromosomes are allelic to those on the *tabacum* chromosome.

Zaitlin: The evidence is against that. When you cross *N. glutinosa* and *N. tabacum*, those chromosomes are incompatible. They are very different.

In addition to the work described in my paper, we have recently shown (unpublished results) that TMV 'H protein' (Asselin & Zaitlin 1978) is a ubiquitinated form of coat protein.

Gianinazzi: Did you detect the H protein only at a shifted temperature or was it detected during systemic infection?

Zaitlin: The H protein had nothing to do with the hypersensitive response. We find it in systemic infections. We actually first found it in the virion population but now we get it from chloroplasts of virus-infected plants.

Lommel: Do you still think that there is a moiety associated with the virion?

Zaitlin: There is some H protein associated with virions. It is possible to do reconstitution experiments and cause H protein to return into the virion. It is protease resistant while it is in the virion.

Lommel: Can you identify its location in the virion?

Zaitlin: We have tried, but we cannot localize it to one region. It seems to be random.

van Kammen: What has to happen before an H protein is produced? It appears that a few rather specific peptides bonds are made.

Zaitlin: A very specific linkage is made between the C-terminal end of the ubiquitin molecule and the ε-amino group of a lysine on the target molecule. There are only two lysines in the coat protein of TMV so it probably has to be one of those, if it is that type of reaction. The lysines are towards the amino end of TMV and are exposed on the surface.

Bruening: Is the carboxyl attachment of ubiquitin consistent with your earlier data on the ends of the H protein?

Zaitlin: Yes, because the ends of the coat protein are intact. We know that the amino end of the ubiquitin molecule is free.

Bruening: Was there earlier evidence that the C-terminal end of the attached protein molecule was free?

Zaitlin: We postulated that H protein could be a ubiquitinated coat protein because of this very unusual isopeptide bond. We also had some amino acid data—when we subtracted the amino acid sequence of the coat protein from the H protein, the remainder looked like ubiquitin (Collmer et al 1983).

Lommel: Is there any possibility of synthesizing this molecule *in vitro*?

Zaitlin: I suppose we could ubiquitinate coat protein. I don't think that would be particularly useful.

Harrison: Perhaps ubiquitination of coat protein may be important in the development of disease symptoms, particularly those where chloroplasts are affected.

van Vloten-Doting: If this phenomenon really has something to do with viral symptoms one might expect that, if one takes another virus which induces chlorosis, ubiquitin will be linked to that particular coat protein in the chloroplasts.

Zaitlin: We have started to look at other viruses to see whether they end up in a chloroplast.

Harrison: Is it your working hypothesis, then, that some of the ubiquitinated TMV coat protein is at the end of the virus particles and facilitates removal of the first few coat protein subunits when the particles become associated with ribosomes in the initial stages of infection?

Zaitlin: I wouldn't go that far. We don't know why it is associated with virions.

Beachy: We also used an antibody against ubiquitin and discovered that ubiquitin was present in virus-infected plants, in contrast to healthy plants. Ubiquitin antibody cross-reacted with chloroplast proteins, which fits the model Dr Zaitlin proposed.

We have been looking at early events in disease development and in chloroplast function, and the association of coat protein in thylakoids is a real one. We find capsid protein inside thylakoids before chlorosis is seen, at the same time that we observe a decrease in electron flow. This interruption in electron flow is between the water-splitting reaction and the cytochrome S. It is not in PSI, but affects the oxidation and reducing sides of PSII. I think Dr Zaitlin has described an important concept in pathogenesis.

Zaitlin: Did you find any reaction with ubiquitin antibodies in uninfected protoplasts?

Beachy: We did not.

Zaitlin: This is worrying. Ubiquitin is not easy to detect on Western blots. A study has been done in which the Western blot was autoclaved before the antibodies were added (Swedlow et al 1986). That enhanced the signal dramatically.

Harrison: What technique were you using to detect ubiquitin in chloroplasts, Dr Beachy?

Beachy: We purified chloroplasts, treated them with protease, and did a

Western blot with ubiquitin antibodies and added secondary ^{125}I label.

Zaitlin: What size was the molecule you detected?

Beachy: One was similar to your H protein in size, but we didn't know what it was and couldn't correlate it with anything. Inside the thylakoids we found the coat protein antibody to cross-react with a number of other peptides, not just those of H protein size. We do not know where those come from yet but they are real and associated with infection. We have, however, located the protein that Dr Zaitlin has found, but couldn't correlate it as we did not have the H protein antibody.

References

Asselin A, Zaitlin M 1978 Characterization of a second protein associated with virions of tobacco mosaic virus. Virology 91:173–181

Boudon E, Martin C, Carré M 1970 Une souche aucuba thermosensible du virus de la mosaique du tabac. C R Acad Sci Paris Ser D Sci Nat 270:2508–2511

Bruening G, Ponz F, Glascock C, Russell ML, Rowhani A, Chay C 1987 Resistance of cowpeas to cowpea mosaic virus and to tobacco ringspot virus. In: Plant resistance to viruses. Wiley, Chichester (Ciba Found Symp 133) p 23–37

Collmer CW, Vogt VM, Zaitlin M 1983 H protein, a minor protein of TMV virions, contains sequences of the viral coat protein. Virology 126:429–448

Dumas E, Gianinazzi S 1986 Pathogenesis-related (b) proteins do not play a central role in TMV localization in *Nicotiana rustica*. Physiol Mol Plant Pathol 28:243–250

Fritig B, Kauffmann S, Dumas B, Geoffroy P, Kopp M, Legrand M 1987 Mechanism of the hypersensitivity reaction of plants. In: Plant resistance to viruses. Wiley, Chichester (Ciba Found Symp 133) p 92–108

Gerstel DU 1945 Inheritance in *Nicotiana tabacum*. XIX. Identification of the *tabacum* chromosome replaced by one from *N. glutinosa* in mosaic-resistant Holmes Samsoun tobacco. Genetics 30:448–454

Kubo S, Oohashi Y, Tomaru K 1971 Ann Phytopathol Soc Jpn 35:402

Melchers G, Jockusch H, Sengbusch PV 1966 A tobacco mutant with a dominant allele for hypersensitivity against some TMV strains. Phytopathol Z 55:86–88

Smart TE, Dunigan DD, Zaitlin M 1987 *In vitro* translation products of mRNAs derived from TMV-infected tobacco exhibiting a hypersensitive response. Virology 158:461–464

Swedlow PS, Findley D, Varshavsky A 1986 Enhancement of immunoblot sensitivity by heating of hydrated filters. Anal Biochem 156:147–153

Mechanisms of cross-protection between plant virus strains

John L. Sherwood

Department of Plant Pathology, Oklahoma State University, Stillwater, Oklahoma 74078–0285, USA

Abstract. Cross-protection (the phenomenon whereby the activity of a virus in a plant prevents the expression of a subsequent challenge virus) has been used successfully to control some virus diseases. Mechanisms to account for specificity have been proposed that operate either at the initial interaction between the plant infected with the protecting virus and the challenge virus, or during the replication of the challenge virus. In the initial interaction, the challenge virus could be inhibited from uncoating, thereby preventing the initiation of the replicative cycle. If replication is initiated, a number of mechanisms may be involved in controlling replication of the challenge virus: (1) the initial translation of the incoming viral nucleic acid could be blocked, (2) the transcription of the incoming viral nucleic acid may be prevented even if it is translated initially, and (3) the production of genome-length viral nucleic acid could be inhibited. Finally, even if challenge virus is replicated, movement from cell to cell could be prevented. Explanation of cross-protection by one hypothesis alone is unlikely because of the contrasting observations in a variety of biological systems. However, it is plausible that different mechanisms of cross-protection may be operating in different virus groups.

1987 Plant resistance to viruses. Wiley, Chichester (Ciba Foundation Symposium 133) p 136–150

The phenomenon of the activity of a virus in a plant preventing the expression of a subsequent challenge virus has received many names (acquired immunity, acquired tolerance, cross-immunization, premunity), but is commonly known as cross-protection. Cross-protection has been used successfully to control some virus diseases, although the specific mechanism of protection remains unexplained. In this paper, several hypotheses addressing the mechanism of cross-protection between plant viruses are reviewed and the possibility that different mechanisms operate in different cross-protection systems is discussed.

Some observations in systems used to study cross-protection

Since cross-protection was first investigated in the early 1900s, many approaches have been used to study the phenomenon. Whole plants, parts of plants, and protoplasts have all been used to study the events in cross-protection. With whole plants, cross-protection has been evaluated in the leaf that received the protective inoculation directly or became infected as a result of the systemic nature of the protective virus. The time interval between the protective inoculation and the challenge virus inoculation has varied considerably among the cross-protection systems studied. Assays to measure cross-protection or superinfection (the lack of complete protection) have also varied with the virus–host system used. Hence, it is difficult to compare results from different cross-protection studies. What has been demonstrated is that: (1) the challenge virus may or may not multiply, (2) cross-protection can be unilateral or bilateral between virus isolates, and (3) the amount of protection between viruses is relative and depends on the conditions of the experiment.

Of particular interest has been the reduced viral content and the reaction to challenge inoculation of dark green areas of mosaic leaves. Atkinson & Matthews (1970) showed that dark green islands in tobacco mosaic virus (TMV)-infected tobacco (*Nicotiana tabacum* L.) maintained a low concentration of virus over a period of weeks, although they were contiguous to light green areas having a high concentration of TMV. The dark green areas were cytologically normal and appeared to maintain normal cytological connections with light green areas. The concentration of the systemically infecting strain could not be increased in the dark green areas by subsequent mechanical inoculation. The authors suggested that dark green areas could not arise as escapes from infection during the ontogeny of the leaf but must develop and be maintained by a 'dark green agent'.

Carlson & Murakishi (1978) followed the ontogeny of clonally derived cells in the leaf using *N. tabacum* var. John Williams Broadleaf with a specific genetic marker that allowed identification of cells that were non-clonally derived, but in physical contact with one another. By regenerating virus-free plants from the non-clonally derived cells in the same dark green areas they concluded that dark green areas did not arise from a single cell and that an 'intercellularly communicable agent is involved in producing green islands'.

Shalla & Peterson (1978) studied the development of insusceptibility of leaves that had been directly inoculated with TMV to superinfection by another strain of TMV. They concluded that 75% of four million potentially infectible cells remained uninfected by the initial virus inoculation. The entire leaf, however, was insusceptible to the challenge virus, suggesting 'there is some substance formed in response to infection which moves into cells surrounding the initial infection centers which somehow inhibits replication of related virus entering those cells'.

Protoplast systems have also been utilized to study cross-protection. Otsuki & Takebe (1976) demonstrated that two related strains of TMV could infect the same protoplast. In successive inoculations, however, the longer the period between the two inoculations the greater the exclusion of the challenger. Although protoplasts lose their susceptibility in culture, the loss of susceptibility of protoplasts infected with one strain of TMV was much greater to infection by another strain of TMV than to cucumber mosaic virus (CMV). Barker & Harrison (1978) found similar results with the RRV-S and RRV-E strains of raspberry ringspot virus (RRV). Sequential inoculation resulted in exclusion of the challenger and the longer the interval between inoculations the greater the effect. Leaves produced on a plant after the protective inoculation with the systemically infecting RRV-S strain did not show symptoms of RRV-E infection after inoculation with the isolate, but protoplasts from leaves of an RRV-S-infected plant could be superinfected by RRV-E. This could indicate either that RRV-E infects the intact leaves but no symptoms result, or that the protoplasts from an RRV-S-infected plant behave in a manner different from the intact plant. This may further support the notion that a diffusible factor is required for the complete expression of cross-protection.

Hypotheses to account for specificity in cross-protection

Infection of a plant with a virus can reduce significantly the susceptibility of the plant to infection by other viruses, regardless of the relationship of the second virus to the virus used for the initial inoculation. The same virus combination in another host may have an entirely different outcome. This suggests that at least two separate phenomena operate in cross-protection. One is that a viral infection may reduce the susceptibility of the plant to subsequent infection by any other virus. The host plant may play an important role in this event. The second is that there must be a mechanism to account for the specific protection against superinfection by related viruses. Several theories based on experimental evidence or speculative insight have been proposed to explain the mechanism underlying specificity in cross-protection.

Specific metabolite utilization

Kohler & Hauschild (1947) proposed that cross-protection results from the utilization of some metabolite(s) by the first virus that is also essential to the second virus. However, in studies with different host and virus combinations, the primary infection generally inhibited the development of the second virus regardless of the morphological or serological properties of the two viruses. Ross (1974) proposed that the resistance-inducing virus would sequester the majority of the ribosomes in a cell through a rapid increase in mRNA. Thus, when the challenge virus was added, were it related or not, few ribosomes would be left for binding and the nucleic acid would be degraded prior to expression. This hypothesis could serve to account for the non-specific resistance that develops in a virus-infected plant to infection by another virus.

Specific binding of viral components

Several proposed mechanisms of cross-protection have centred on the possibility that the challenge virus, upon introduction, could be adsorbed to virus aggregates or virus components, thereby becoming inactivated. Gibbs (1969) suggested that in a cross-protection test the nucleic acid of the challenge virus could irreversibly, but ineffectually, bind to the replicase of the virus already present in the cell if the challenge virus had a similar, but not identical, replicase recognition site. However, a viral nucleic acid with an identical or quite different replicase recognition site could replicate and be translated unhindered. The sequence and secondary structure of the 3' end of the viral nucleic acid, where the initial interaction between the virus and replicase occurs, is generally conserved within a virus group. In contrast, Ross (1974) suggested that the added challenger viral RNA is as readily recognized and copied by the existing replicase as it is by its own. Therefore, synthesis of more replicase from the RNA of the challenger would give the challenge virus no advantage over the preexisting virus since no new enzyme capability would have been added. The only new addition would be that of the challenger template. In all probability the amount of effective new template added to any cell by inoculation would be extremely small compared to that of the preexisting virus. The production of pseudorecombinants of RRV (Harrison et al 1974) and other viruses demonstrates that related strains of a virus can utilize the same replicase. Additionally, the larger RNAs of CMV and tobacco streak virus (TSV) have been implicated in encoding the replicase function and in cross-protection. Hence, replicase may in some way be involved in cross-protection.

De Zoeten & Fulton (1975) proposed that the molecular basis for cross-protection is the elimination of the genome of a superinfecting related virus by its capture in the coat protein of the virus of the original infection. Zaitlin (1976) tested their hypothesis using a coat protein mutant of TMV (PM1) that produces insoluble coat protein that does not encapsidate TMV RNA. When plants inoculated with the mutant were challenge-inoculated with TMV (U1), a lower amount of the U1 strain was recovered than in controls, indicating that protection was achieved.

Palukaitis & Zaitlin (1984) proposed that for positive-sense (+)RNA viruses, superinfection by the challenge virus would be prevented by the inhibition of synthesis of the (+)RNA of the challenge virus. Inhibition of synthesis would occur because the negative-sense (−)RNA produced from the initial infectious RNA of the challenge virus would be bound by (+) RNA of the protecting strain. Antisense RNA has been shown to have an effect on gene regulation in both prokaryotic and eukaryotic systems. The satellite RNA of CMV, when in an antisense sequence, will specifically bind *in vitro* to residues 557–589 of the coat protein gene of the Q-strain of CMV and may be involved in the regulation of coat protein production (Rezaian & Symons 1986). However, the involvement of antisense RNA would require every cell of the protected plant to contain enough antisense RNA to bind to any RNA from the challenge inoculation that entered the cell.

Inhibition of systemic spread

Early research suggested that the basis of cross-protection was the inability of the challenge virus to replicate and spread through the plant. Methods using differences in lesion size or differences in antigenicity have been used to analyse the replication and systemic spread of virus strains in cross-protection experiments. Dodds (1982) and Dodds et al (1985), however, were able to follow the multiplication of two strains of CMV directly, because of differences in the electrophoretic mobility of their virions and double-stranded (ds) RNAs. When virion accumulation was followed in reciprocal cross-protection experiments between a mild and severe strain of CMV, each strain protected against systemic infection by the other. It was only when the virus from the initial inoculation was not well established that the challenge strain could increase in the leaves that received the challenge inoculation of the protected plant. However, the challenge strain was not able to move away from this site of initial multiplication.

Involvement of viral coat protein

Sherwood & Fulton (1982) proposed that the specific basis of cross-protection with TMV in *N. sylvestris* is the inability of the challenge virion to uncoat when inoculated onto plants systemically infected with the protecting strain. If a mosaic leaf is inoculated with a strain of TMV that produces necrotic lesions, the lesions are restricted to the dark green areas of the mosaic. However, if mosaic leaves are inoculated with RNA from a necrotic lesion-producing TMV strain, lesions are produced in both light and dark green areas of the mosaic. The specificity in the inhibition of uncoating may involve the kind of coat protein present in the protected cell. RNA of a necrotizing strain of TMV, encapsidated in brome mosaic virus protein and used as a challenge, superinfected in the same manner as RNA. However, RNA of a necrotizing strain of TMV encapsidated in coat protein of the common strain was as unable to superinfect as was native virus (Sherwood & Fulton 1982). Two to 7.5 times as much TMV antigen can be detected by ^{125}I-labelled TMV-specific IgG in cell walls of light green tissue of mosaic leaves as in dark green tissue (De Zoeten & Gaard 1984). This may indicate that necrotic lesion-producing strains do not produce necrotic lesions in the light green areas because of the inability of intact virions to uncoat in tissue infected with a high concentration of the common strain of TMV.

Dodds et al (1985) found similar results with a tomato–CMV system. In leaves infected with the protecting strain of CMV, a breakdown of cross-protection occurred when the protected leaves were challenge-inoculated with viral RNA but not with virus particles. In *in vitro* experiments, Wilson & Watkins (1986) found that the uncoating and subsequent translation of TMV in rabbit reticulocyte lysate could be inhibited by adding TMV coat protein. Experiments in the *N. sylvestris* system with a mutant of TMV that lacks coat protein support the involvement of coat protein in the cross-protection phenomenon (Sherwood 1987). Leaves of *N. sylvestris* inoculated previously with the mutant or the common strain of TMV were challenged with either turnip mosaic virus (TuMV), to test the non-specific protection induced by the mutant, or a necrotizing strain of TMV. An equal reduction in susceptibility to both TuMV and the necrotizing strain occurred in leaves inoculated with the mutant, but in leaves inoculated with the common strain there was a greater reduction in susceptibility to the necrotizing strain than to TuMV. Work with the same mutant in a different host, however, gave different results (Sarkar & Smitamana 1981). In *Nicotiana tabacum* cv. Samsun EN, the coat protein mutant significantly reduced the number of necrotic lesions produced to a challenge strain of TMV as compared to control leaves.

The hypothesis that coat protein may be involved in cross-protection may be supported by the work of Jockusch (1968). He demonstrated that TMV strain Ni2519, which produced no detectable protein when kept at 35 °C, would not protect the plant from subsequent infection with a vulgare strain. This was in contrast to plants infected with Ni118, a strain which does produce protein at 35 °C, but the protein does not assemble with its own nucleic acid and cross-protection against the vulgare strain is observed. In fact, more TMV coat protein synthesis occurs with the Ni118 at 35 °C than at 23 °C. The coat protein produced by the Ni118 mutant at 35 °C is pelleted out by centrifugation at 9000 g for 20 min, which may indicate that the coat protein produced is associated with large particles in the cell, such as the cell wall, where the virus may first encounter its uncoating environment.

Cross-protection studies with the tobamoviruses, sunn hemp mosaic virus (SHMV) and TMV, indicate that coat protein may be a factor in cross-protection in some situations, but other factors may be involved (Zinnen & Fulton 1986). SHMV protected completely against superinfection by either virions or RNA of a necrotizing strain of SHMV produced by nitrous acid treatment. When plants infected with SHMV were challenged with RNA of TMV encapsidated in either SHMV coat protein or TMV coat protein, the RNA encapsidated in SHMV coat protein was 5–27 times less infectious than the TMV RNA encapsidated in TMV coat protein.

The development of plant transformation systems has allowed specific investigation of the effect of the expression of the viral coat protein gene in the reaction of a plant to inoculation with a virus. Powell Abel et al (1986) developed transgenic tobacco plants which express the TMV coat protein. Transformed plants did not develop symptoms after inoculation at very low test virus concentrations. When the test inoculum was increased, the percentage of plants that became infected increased and the delay of symptoms seen at the lower challenge inoculum concentration was reduced. These observations are consistent with the hypothesis that the number of infections are reduced in leaves of transgenic plants that express the coat protein gene, compared with leaves of control plants. Loesch-Fries et al (1987) were able to achieve expression of alfalfa mosaic virus (AlMV) coat protein in tobacco transformed with a Ti plasmid vector carrying AlMV 4 cDNA. Protoplasts isolated from transformed plants could be infected by the genomic RNAs of AlMV. This indicates that the AlMV coat protein produced was biologically active, since it has been shown that RNA-4, or the coat protein it codes for, is required for the infection by the genomic RNAs. When protoplasts from transformed plants were challenged with AlMV virions, the protoplasts were protected from infection. Although coat protein may be specifically involved in some systems of cross-protection, there are many examples where serologically and biologically related viruses do not cross protect (e.g. TSV). Therefore, other factors must be involved, either by themselves or together with coat protein.

Is there a single explanation?

It is clearly impossible to reconcile a single hypothesis for the mechanism of cross-protection with the breadth of the observations made in biological systems. There may be several events that affect the interaction of the protecting and challenge viruses. The first of these is non-specific resistance. The infection of a host by a virus can reduce the susceptibility of the host to infection by a challenge virus, regardless of whether the challenge virus is related to the protecting virus. This effect may also be dependent on the host.

Several mechanisms may be involved in the specific protection that can occur between two viruses. Mechanisms to account for specificity could operate either at the initial interaction between the plant infected with the protecting strain and the challenge virus, or during the replication of the challenge virus. In the initial interaction the challenge virus could be inhibited from uncoating, thereby preventing the initiation of the replicative cycle. If the replication is initiated, a number of mechanisms may be involved in the control of the replication of a challenge RNA virus: (1) the initial translation of the incoming (+)RNA could be blocked, (2) the transcription of the incoming (+)RNA could be prevented even if it is initially translated, and (3) the production of nascent (+)RNA strands from a (−)RNA could be inhibited. Many of the mechanisms proposed could account for these events. Finally, if challenge virus is replicated it could be prevented from moving cell to cell. Perhaps translation of specific proteins required for cell-to-cell movement could be blocked. The observations invoking a 'dark green agent' in cross-protection support the idea that control of cell-to-cell movement is involved as a mechanism to account for specificity in protection. Different mechanisms could operate in different groups of viruses. Coat protein seems to be involved in cross-protection with TMV, CMV and AlMV. All three viruses have been grouped into a 'Sindbis-like' virus group (Goldbach 1986). Elucidation of the mechanisms involved in viral gene expression and regulation with plant transformation systems may reveal which, if any, of these mechanisms are involved in plant viral cross-protection.

Acknowledgements

Thanks are due to G. A. DeZoeten, J. A. Dodds, R. W. Fulton, L. S. Loesch-Fries, U. K. Melcher and T. M. Zinnen for review of this manuscript. This work supported in part by NSF grant NSF–RII–OK–8610676–7 and the Oklahoma Agricultural Experiment Station. Journal Article No. 5158, Oklahoma Agricultural Experiment Station, Oklahoma State University.

References

Atkinson PH, Matthews REF 1970 On the origin of dark green tissue in tobacco leaves infected with tobacco mosaic virus. Virology 40:344–356

Barker H, Harrison BD 1978 Double infection, interference and superinfection in protoplasts exposed to two strains of raspberry ringspot virus. J Gen Virol 40:647–658

Carlson PS, Murakishi HH 1978 Evidence of the clonal versus non-clonal origin of dark green islands in virus infected tobacco leaves. Plant Sci Lett 13:377–381

De Zoeten GA, Fulton RW 1975 Understanding generates possibilities. Phytopathology 65:221–222

De Zoeten GA, Gaard G 1984 The presence of viral antigen in the apoplast of systemically virus-infected plants. Virus Res 1:713–725

Dodds JA 1982 Cross-protection and interference between electrophoretically distinct strains of cucumber mosaic virus in tomato. Virology 118:235–240

Dodds JA, Lee SQ, Tiffany M 1985 Cross protection between strains of cucumber mosaic virus: effect of host and type of inoculum on accumulation of virions and double stranded RNA of the challenge strain. Virology 4:301–309

Gibbs AJ 1969 Plant virus classification. Adv Virus Res 14:263–328

Goldbach RW 1986 Molecular evolution of plant RNA viruses. Annu Rev Phytopathol 24:289–310

Harrison BD, Murant AF, Mayo MA, Roberts IM 1974 Distribution of determinants for symptom production, host range and nematode transmissibility between the two RNA components of raspberry ringspot virus. J Gen Virol 22:233–247

Jockusch H 1968 Two mutants of tobacco mosaic virus temperature-sensitive in two different functions. Virology 35:94–101

Kohler E, Hauschild I 1947 Betrachtungen und Versuche zum Problem der 'erworbenen Immunität' gegen Virusinfektionem bei Pflanzen. Zuchter 17–18:97–105; abstr in Rev Appl Mycol 29:78

Loesch-Fries S, Halk E, Merlo D et al 1987 Expression of alfalfa mosaic virus coat protein gene and anti-sense cDNA in transformed tobacco tissue. UCLA (Univ Calif Los Ang) Symp Mol Cell Biol 48:221–234

Otsuki Y, Takebe I 1976 Double infection of isolated tobacco leaf protoplasts by two strains of tobacco mosaic. In: Tomiyama JM et al (eds) Biochemistry and cytology of plant parasite interaction. Elsevier, Amsterdam, p 213–222

Palukaitis P, Zaitlin M 1984 A model to explain the 'cross-protection' phenomenon shown by plant viruses and viroids. In: Kosuge T, Nester EW (eds) Plant–microbe interactions: molecular and genetic perspectives. Macmillan, New York, p 420–429

Powell Abel P, Nelson RS, De B et al 1986 Delay of disease development in transgenic plants that express the tobacco mosaic virus coat protein gene. Science (Wash DC) 232:738–743

Rezaian MA, Symons RM 1986 Anti-sense regions for satellite RNA of cucumber mosaic virus form stable complexes with the viral coat protein gene. Nucl Acids Res 14:3229–3239

Ross AF 1974 Interaction of viruses in the host. In: Lawson RH, Corbett MK (eds) Virus diseases of ornamental plants. International Society of Horticultural Science, The Hague, Netherlands, p 247–260

Sarkar S, Smitamana P 1981 A proteinless mutant of tobacco mosaic virus: evidence against the role of a viral coat protein for interference. Mol Gen Genet 184:158–159

Shalla TA, Peterson LJ 1978 Studies on the mechanism of viral cross protection. Phytopathology 68:1681–1683

Sherwood JL 1987 Demonstration of the specific involvement of coat protein in tobacco mosaic virus (TMV) cross protection using a TMV coat protein mutant. J Phytopathol 118:358–362

Sherwood JL, Fulton RW 1982 The specific involvement of coat protein in tobacco mosaic virus cross protection. Virology 119:150–158

Wilson TMA, Watkins PAC 1986 Influence of exogenous viral coat protein on cotranslational disassembly of tobacco mosaic virus (TMV) particles *in vitro*. Virology 149:132–135

Zaitlin M 1976 Viral cross-protection: more understanding is needed. Phytopathology 66:382–383

Zinnen TM, Fulton RW 1986 Cross protection between sun-hemp mosaic and tobacco mosaic viruses. J Gen Virol 67:1679–1687

DISCUSSION

van Vloten-Doting: You describe several mutants of TMV that produce coat protein which does not encapsidate TMV RNA. For alfalfa mosaic virus (AlMV) we know that the coat protein has two different functions: an early function (van Vloten-Doting 1977), probably in replication (Houwing & Jaspars 1978), and a late function in encapsidation. The two functions can be mutated separately (van Vloten-Doting 1985). Perhaps the coat protein in other viruses has additional roles besides being a protective coat. The different mutations may affect these different functions.

Sherwood: That is quite right.

Dodds: Dr Sherwood, all the experiments on interactions of tobacco strains that you describe involve local lesion-producing strains as the challenge virus. Protection can therefore be detected by counting the number of local lesions produced. What about studies in which the challenge strain is expected to be systemic in the host? In these cases symptom expression will not be visible if cross-protection is active, but does TMV accumulate in these plants? The study by Cassells & Herrick (1977) appeared to demonstrate the accumulation of TMV in protected tomato. Have other similar studies been made with TMV?

Sherwood: We have tried superinfecting the dark green areas of *Nicotiana sylvestris* (resulting from systemic infection by the common strain of TMV) with high concentrations of the RNA of the same strain of TMV. RNA is a more efficient inoculum than the virion when one challenge-inoculates with a strain of TMV that produces necrotic lesions on *N. sylvestris*. We did not observe an increase of the common strain in the dark green areas that were challenged. This is therefore one example of the inhibition of TMV multiplication by the same strain.

Dodds: Cassells & Herrick (1977) found nearly as much challenge-strain virus throughout the tomato plant as would have been expected in the absence of any protection. The only distinction was the lack of symptom expression by that virus. Is it a true generalization that TMV accumulates in protected plants?

Harrison: I do not know for TMV. However, we have worked on a rather different system—raspberry ringspot virus (RRV) in *N. benthamiana* (Barker & Harrison 1978 and unpublished results). Fluorescent antibody staining of protoplasts from recovered RRV-infected leaves that were not showing symptoms demonstrated that every cell contained virus antigen. This distinguishes RRV from the mosaic-inducing viruses. When these RRV-infected leaves were challenge-inoculated with another strain of RRV no symptoms were produced but we found, by making protoplasts and using strain-specific antisera, that a large proportion of the cells were infected with the challenge strain: 17% when plants infected with RRV strain E were challenged with RRV-S, and 74% when those infected with RRV-S were challenged with RRV-E.

Dr Sherwood referred to our experiments with pseudorecombinants of RRV. We used virus isolates that have both RNA species from strain E (= EE), both from strain S (= SS), or either heterologous combination, namely RNA-1 from E and RNA-2 from S (ES), or the converse (SE). We showed that it is not necessary to have both parts of the genome of the challenge virus different from those in the protecting virus for superinfection to occur. However, both RNA-1 and RNA-2 are involved in superinfection.

We challenged ES-infected leaves with EE or with particles that contained only RNA-2 of E (−E). There was no additional virus multiplication after inoculation with −E but on inoculation with EE we obtained some infection, but less than in unprotected plants. Assays done by immunosorbent electron microscopy gave the same results as those obtained by fluorescent antibody staining. Putting in the RNA-2 of strain E alone is therefore not enough to induce renewed virus replication, whereas inoculation with EE, which differs only in having the RNA-1 of strain E, which is already in these plants, is effective.

There is an interesting conclusion to be drawn from this, but I am not quite sure what it is! Perhaps there is some kind of compartmentation within the cell, so that an inoculum of the challenge virus must contain RNA-1 if the challenge virus is to multiply even when the identical RNA-1 already exists in the plant cell. In nepoviruses, the group to which RRV belongs, RNA-1 can replicate in cells on its own but RNA-2 is only replicated in cells that contain RNA-1 (Robinson et al 1980). Although this nepovirus system is rather different from the ones we have been discussing, the results suggest that cross-protection mechanisms are not always simple.

Zaitlin: Are there any examples of intact plant protection in which RNA is effective where virions are not?

Dodds: There would not be a situation where RNA could be used as a natural inoculum.

Zaitlin: I realize that, but I refer to whole plant protection in which one challenges with either virus or RNA, rather than these more artificial studies on local lesions.

Dodds: Dr Sherwood mentioned my work with CMV. In a protected leaf that was challenged with virions, there was no accumulation of the challenge virus, whereas when CMV RNA was used for the challenge in the CMV-infected leaf, the challenge virus did accumulate. However, there was no subsequent invasion of the rest of the plant. It is necessary to distinguish between events in the inoculated leaf and in the rest of the plant.

Harrison: Perhaps cells in the tip leaves were already occupied by the protecting strain by the time the challenge strain got to them.

Dodds: That is possible, but the distinction I was trying to make is that when the experiment is done with TMV, even though no symptoms are expressed, TMV does accumulate in the rest of the plant.

Harrison: In some experiments with nepoviruses one can recover the challenge strain from the tip leaves (Murant et al 1968).

Zaitlin: Therefore, in some sense cross-protection becomes a matter of inhibition of symptoms.

Dodds: However, for cucumber mosaic virus, there can be a complete blockage of any challenge virus activity throughout the whole plant.

Harrison: Both the inhibition of symptoms and blockage of virus replication seem to be aspects of cross-protection with different relative expressions in different systems.

Fraser: We have looked at the multiplication of different TMV isolates using the same system of coat proteins of different charge which can be separated in 8M urea gels. We found that the challenging isolate will spread through the protected plant but on average its multiplication is inhibited by up to 90%. Therefore, I don't think we are just looking at prevention of symptoms: we are looking at a strong effect on multiplication.

Dodds: Are these results different from those of Cassells & Herrick (1977)?

Fraser: I would not like to over-interpret the results because they are based on differences in lesion size and serology which are difficult to quantify.

Davies: Are there any mutants that are derived from a strain of challenge virus which was protected, but have reverted to not being protected or have even become synergistic? An analysis of such mutants might be an approach to understanding the mechanism and asking whether uncoating, translation, transcription or cell-to-cell movement are involved in cross-protection.

Sherwood: I don't know if anyone has tried to do that. The fact that some isolates of tobacco streak virus only unilaterally cross-protect would indicate that it is a possibility (Fulton 1978). I don't know what the difference is among those isolates. That might be interesting to study.

Beachy: What is the relative amount of coat protein in the green islands compared to the yellow islands?

Sherwood: De Zoeten & Gaard (1984) found about 2–7.5 times as much coat protein in the light green islands as in the dark green islands. That is quite a large amount of coat protein when one compares it with the amounts of virus in the light and dark green areas—there are many virus-free cells in the dark green areas but approaching 1 mg per gram of tissue of virus in the light green areas.

Harrison: What is the evidence that the dark green island cells are virus-free?

Matthews: There is no conclusive proof for this. However, in cells on the infected side of the boundary of the dark green areas, typical crystals of TMV were visible. On the other side, some cells away from the boundary, no crystals of rods or single rods were visible at all, but there was a gradient over a zone of about four or five cells in which the number of virus rods seen decreased to zero. We interpreted that to be due to diffusion of virus from infected cells into the dark green cells where it could not replicate. It is not possible to prove that there is no virus at all in the dark green areas.

Sherwood: Murakishi & Carlson (1976) were able to regenerate virus-free plants from dark green areas. I don't know whether the virus was lost during the regeneration, rather than not being present at the start.

Harrison: Virus-free plants can be generated from infected meristems too.

Matthews: If one is doing analysis of any sort on dark green islands, it is difficult to be sure that microscopic clusters of infected cells are not present. In the system we have studied most, turnip yellow mosaic virus in Chinese cabbage, one has a clear cytological marker of infection—the virus-induced vesicles on the outer chloroplast membrane. When we cut hand sections we found microscopic islands of one or more cells showing cytological indicators of full infection, even in the areas that were apparently pure dark green to the naked eye.

Loebenstein: On the other hand, we prepared protoplasts from CMV green islands with Dr Roger Wood. We used fluorescent antibody staining and could not detect any coat protein or virus. Extracting virus from these protoplasts gave only very minute levels, if any, of infectivity.

Matthews: For many years I have been interested in the resistance of dark green islands in virus-infected leaves showing mosaic disease, and in the mechanism by which these islands of tissue arise during leaf ontogeny. Unfortunately, it is difficult to apply molecular biological techniques to this problem. The detailed pattern of dark green tissue in each mature leaf is different and irregular. If one goes back to a small leaf initial in the apical dome region, where the crucial events must take place, one cannot predict which parts of the leaf initial would have given rise to dark green islands in the mature leaf.

A graduate student, Ms Allayne Ferguson, and I have recently defined a system in which it may be possible to apply certain molecular biological techniques. We inoculated the cotyledon leaves of very small Chinese cabbage plants with turnip yellow mosaic virus when there were only about three or four true leaf initials. As the plants grew and became systemically infected, the fifth leaf to emerge developed a clearly defined ring of dark green tissue around the entire leaf margin in about two-thirds of the plants. We were able to establish that this fifth leaf was about 0.6 mm long at the time it became infected. At this stage midrib and lamina cells could be distinguished. The cells in the marginal zone giving rise to the dark green zone in the mature leaf were still in a fully meristematic state, whereas those in the central zone were vacuolated and were infected. Thus in this system, at least, it appears that there is some property of cells that are still fully meristematic when the leaf is invaded by virus which triggers them to become dark green rather than normally infected. With techniques such as immuno-gold staining for viral proteins and *in situ* hybridization for viral nucleic acids, it might be possible to investigate a time sequence of events in this fifth leaf with at least a two out of three chance that one could correctly predict the pattern of dark green tissue in the mature leaf.

Fraser: The location of dark green tissue around the veins is a common feature of dark green islands. We have been interested in possible hormonal explanations of where the dark green tissue comes from, why it is darker green, and why it is resistant to virus.

We have used TMV-infected N' gene tobacco so that we could have systemic or local lesion reactions. We used the local lesion reaction to test the sensitivity of tissue to inoculation. The chlorophyll concentration in dark green areas is typically twice that in light green areas of infected leaves, or of healthy controls. The TMV concentration has been assessed by counting particles by electron microscopy. In the dark green areas it is less than 1% of the level in light green areas.

Abscisic acid (ABA) is a plant stress hormone, and in sham-inoculated controls it is typically found at a level of about 10 ng per gram. There is a statistically significant increase in the light green infected tissue, normally about two- to fivefold, depending on the experiment. In the dark green tissue, which is virtually virus-free, ABA concentration is typically up by five- to tenfold. In addition, ABA forms various stable metabolites, such as glucosyl esters, and there is an oxidative pathway to phaseic acid and dihydrophaseic acid. The latter form glucosyl esters, and so on. If one adds up all these metabolites, one also finds higher levels of the metabolites in the light green, but especially in the dark green, tissue than in control leaves. Therefore we are not only looking at a higher steady state concentration of ABA, but also at a considerably increased rate of synthesis and flux through the pathway.

Is this increased level of ABA of any importance in formation of the dark green areas and in the resistance of this dark green tissue to infection? On the face of it a four or five-fold increase in ABA is not very much, especially in hormone terms where we would be looking for a change of a couple of orders of magnitude. In an unstressed cell, ABA is normally sequestered within the chloroplast and water stress allows it to escape to the cytoplasm where further ABA synthesis is depressed. In infected tissue we find that the increased ABA is entirely outside the chloroplast, and this applies in the dark green as well as in the light green tissue. In terms of extra-chloroplastic ABA there is about a hundred times more in the dark green tissues than in control leaves. That is a good amplification which would fit a typical dose–response curve for a hormone effect.

Does ABA do anything outside the chloroplast? If we spray a healthy plant with ABA, the detoxification mechanisms begin to work and phaseic and dihydrophaseic acids and glucosyl esters are formed which are inactive. Therefore the plant must be sprayed every day with ABA; and then a steady-state increased ABA concentration of about five times normal can be maintained if the correct concentrations are used. As far as we know, that ABA is outside the chloroplasts. The chlorophyll concentration of these healthy plants is about twice the normal level. Therefore that extra ABA is increasing the chlorophyll concentration. If one challenge- inoculates these plants with a local lesion strain of virus, between a tenth and a third as many lesions are formed. Therefore it appears that the artificially added ABA is both making the plants darker green and causing a degree of resistance to infection by virus.

If one inoculates a lower leaf, ABA synthesis is increased there. Using radioactive ABA tracer one can follow the transport of ABA up into the higher leaves. My hypothesis is that there is transport of ABA in the veinal system to the upper leaves where it causes dark green areas around the veins. It may also cause changes in development of these cells, but we have recent evidence suggesting that virus-induced changes in cytokinins may also be involved.

Fritig: Has anybody looked for PR proteins in the dark green islands?

Sherwood: I have never found any in *N. sylvestris* systemically infected with TMV, using Coomassie blue staining (Sherwood 1985).

Fraser: We have looked in ABA-sprayed healthy plants. No PR proteins were detectable, but the plants formed fewer lesions than the controls when inoculated with TMV.

Maramorosch: Has anybody tried to dissociate the green areas in plant tissue culture and grow virus-free plants?

Sherwood: Murakishi & Carlson (1976) regenerated plants from dark green areas of tobacco leaves systemically infected with TMV. The plants were virus-free and were found to be resistant to TMV infection for a couple of months.

Dodds: Although most plants regained susceptibility, one or two tomatoes were still very resistant to TMV (Barden & Murakishi 1985). Dr Murakishi (personal communication) has studied the genetics of resistance in these plants and it seems to be inherited as a cytoplasmic trait.

van Vloten-Doting: It may be some plant variation.

Dodds: That is possible.

References

Barden KA, Murakishi HH 1985 Leaf regeneration and virus screening of somaclones of isogenic lines of tomato. Phytopathol 75:1359 (abstr)

Barker H, Harrison BD 1978 Double infection, interference and superinfection in protoplasts exposed to two strains of raspberry ringspot virus. J Gen Virol 40:647–658

Cassells AC, Herrick CC 1977 Cross protection between severe and mild strains of tobacco mosaic virus in doubly infected tomato plants. Virology 78:253–260

De Zoeten GA, Gaard G 1984 The presence of viral antigen in the apoplast of systemically virus-infected plants. Virus Res 1:713–725

Fulton RW 1978 Superinfection by strains of tobacco streak virus. Virology 85:1–8

Houwing CJ, Jaspars EMJ 1978 Coat protein binds to the 3' terminal part of RNA4 of alfalfa mosaic virus. Biochemistry 17:2927–2933

Murakishi HH, Carlson PS 1976 Regeneration of virus-free plants from dark-green islands of tobacco mosaic virus-infected tobacco leaves. Phytopathology 66:931–932

Murant AF, Taylor CE, Chambers J 1968 Properties, relationships and transmission of a strain of raspberry ringspot virus infecting raspberry cultivars immune to the common Scottish strain. Ann Appl Biol 61:175–186

Robinson DJ, Barker H, Harrison BD, Mayo MA 1980 Replication of RNA-1 of tomato black ring virus independently of RNA-2. J Gen Virol 51:317–326

Sherwood JL 1985 The association of 'pathogenesis' proteins with viral induced necrosis in *Nicotiana sylvestris*. Phytopathol Z 112:48–55

van Vloten-Doting L 1977 Early events in the infection of tobacco with alfalfa mosaic virus. J Gen Virol 41:649–652

van Vloten-Doting L 1985 Virus genetics. In: Francki RIB (ed) The plant viruses, vol 1. Plenum, New York

Genetic engineering of plants for protection against virus diseases

R.N. Beachy*, P. Powell Abel*, R.S. Nelson*, J. Register*, N. Tumer‡ and R.T. Fraley‡

*Department of Biology, Washington University, St Louis, Missouri 63130 and
‡ Monsanto Company, 700 Chesterfield Village Parkway, St Louis, Missouri 63198, USA

Abstract. The practice of protecting plants against severe virus disease by cross-protection has been in use in agriculture for more than 15 years but the molecular mechanism(s) that lead to cross-protection have not been characterized. Tobacco and tomato plants have been produced with some of the characteristics of cross-protection against tobacco mosaic virus (TMV) and alfalfa mosaic virus (AlMV). Transgenic plants were generated that express the coat protein-coding sequence of TMV or AlMV. Plants expressing the TMV or AlMV coat protein were protected against TMV or AlMV infection, respectively, due to an 80–90% reduction in the numbers of sites at which infection is established on the inoculated leaves. Many of the inoculated plants escaped infection; others became infected but expressed a delay in the development of disease similar to the delay in plants that are cross-protected in the classical sense. Protection was also expressed in protoplasts isolated from transgenic plants. In these experiments TMV RNA or TMV was introduced into isolated protoplasts by electroporation. Expression of the TMV (U1 strain) coat protein-coding sequence in transgenic plants protected against a variety of tomato mosaic virus strains (L-TMV, ToMV-1, ToMV-2, ToMV-2a) as well as against the U1 strain and a severe, yellow strain of TMV. Protection against the C_c strain of TMV (sunn hemp mosaic virus) was less dramatic than against other TMV strains. The results suggest that this approach may be used for controlling virus infections when other types of disease resistance genes are not available.

1987 Plant resistance to viruses. Wiley, Chichester (Ciba Foundation Symposium 133) p 151–169

The features of resistance to plant virus disease conferred by different genes/ loci are described elsewhere in this symposium, and further chapters describe novel host gene expression induced in plants in response to virus infections which may in part be responsible for host defence. For more than fifty years a third approach to the protection of plants against some types of severe virus diseases has been recognized — 'cross-protection'. This is an operational term which describes what happens when infection of a plant by a virus protects that plant from superinfection by a second, related strain of the virus. Generally the first virus (the protecting strain) is attenuated for disease symptoms and protects against a severe disease-causing virus (the challenger). As described by Sherwood (this volume), a number of mechanisms have been proposed to explain how cross-protection is effected. Several characteristics of this phenomenon have been described. Some of these apply to most types of cross-protection; others are unique to specific virus groups. It is therefore likely that cross-protection is the result of a series of processes, rather than a single process.

Although neither the precise mechanism responsible for cross-protection, nor the inducing molecule (presumably a product of the infection and/or replication of the protecting strain), has been identified for any cross-protection system, it was predicted that, once identified, the inducing molecule might be expressed in genetically engineered (transgenic) plants in order to produce plants protected against virus infection (Hamilton 1980). This was first successfully achieved in transgenic tobacco plants protected against tobacco mosaic virus (TMV) by Powell Abel et al (1986). Subsequently others reported similar results with alfalfa mosaic virus (AlMV) (Tumer et al 1987). Here we summarize the salient features of these and other unpublished experiments in an attempt to identify the mechanism(s) responsible for the observed protection, and its relationship to classical cross-protection.

Materials and methods: transgenic plants and protection

The methods used to produce transgenic plants that express foreign genes have been described (Fraley et al 1986, Rogers et al 1986). The procedures for cloning cDNAs are common and have been reported elsewhere (e.g. Maniatis 1982). We produced cloned cDNAs that represent the coat protein-coding sequences of TMV and AlMV (Beachy et al 1987a). The complete sequence of the U1 strain of TMV RNA was derived by Goelet et al (1982) and the sequence of alfalfa mosaic virus RNA-3 (a subgenomic of which encodes the coat protein) was reported by Loesch-Fries et al (1985). Specific oligonucleotides were used to prime the cDNA reactions and known restriction sites were used to excise the coat protein-coding sequences of each of the cloned double-stranded cDNAs. These cDNAs were then ligated to the transcriptional promoter for the 35S RNA isolated from cauliflower mosaic virus (CaMV). This promoter is known to be a strong and constitutive promoter in many transgenic plants (Rogers et al 1985, Nagy et al 1985). The 3' ends of the viral cDNAs were ligated with a 3' polyadenylation regulatory sequence isolated from the nopaline synthase gene from the Ti plasmid of *Agrobacterium tumefaciens*. The chimeric gene was contained within an intermediate plasmid which was used to transport the chimeric gene onto a disarmed Ti plasmid in *A. tumefaciens* (Fraley et al 1985). The modified *A. tumefaciens* was used to transform cells of tobacco (*Nicotiana tabacum* cv. Xanthi, Xanthi nc, or Samsun) or tomato leaves (*Lycopersicon esculentum* cv. VF36) and the transformed cells were selected on the basis of their resistance to the antibiotic kanamycin, a trait conferred by the transformation reaction. Transformed cells were regenerated to whole plants by standard procedures (Horsch et al 1985). Regenerated plants were assayed for the presence of the inserted genes and their expression by measurement of the accumulation of messenger RNA and/or protein product.

Transgenic plants that expressed the viral coat protein genes were taken to maturity; the flowers were self-pollinated, and seeds collected. Seedlings germinated from these seeds expressed the gene as a dominant Mendelian trait. Seedlings were grouped into those that expressed the coat protein gene (+CP) and those that did not (−CP). The +CP and −CP seedlings were inoculated with virus and development of disease was monitored in growth chambers or greenhouses. Disease was evaluated by a standardized scale in which the expression of disease symptoms was characterized by vein clearing and a mild chlorosis. Virus replication was assayed by semi-quantitative dot-blot immunoreactions.

Results and discussion

Powell Abel et al (1986) first reported that, in those transgenic Xanthi tobacco plants which express the coat protein-coding sequence of TMV, the accumulation of coat protein accounted for 0.05–0.1% of the total leaf protein. Most of the eight different transgenic plants contained a single copy of the inserted gene, although several contained multiple inserts of the gene, usually at different chromosomal locations. Seedlings produced from parental transgenic plants looked completely normal and developed in the same way as non-transgenic plants. However, when transgenic seedlings that expressed the TMV coat protein gene were inoculated with TMV (0.01–5 µg/ml of the common U1 strain of TMV), symptoms of the infection, if they developed at all, were delayed relative to the development of symptoms in control plants. This was observed with the progeny of four different transgenic lines and was strictly correlated with expression of the coat protein-coding sequence.

Nelson et al (1987) have reported the results of further experiments with these transgenic plants. A severe strain of TMV (ATCC strain PV-230), which induces yellow chlorotic lesions on inoculated leaves and chlorosis on systemically infected leaves, was used. It was observed that there were fewer chlorotic lesions on inoculated leaves of transgenic plants than on control plants. Virus replication (as measured by accumulation of viral antigen) was markedly lower in the inoculated leaves of transgenic plants than in controls, as was virus accumulation in the upper leaves. These results support the hypothesis that expression of the coat protein-coding sequence reduces the numbers of sites at which infection takes place on transgenic plants. They also suggest that, once infection is established, both virus replication in the first leaf and the spread to upper leaves are reduced. This observation was extended to transgenic Xanthi nc (a local lesion host for TMV) plants. When seedlings of such plants were inoculated with U1 TMV, a 95% reduction in the numbers of necrotic local lesions compared to control plants was observed. These results corroborate the results obtained with PV-230 inoculated on the systemic host.

Further characterization of the protection reaction was provided when leaves of transgenic plants were inoculated with TMV RNA rather than virions. When RNA was used as the inoculum the numbers of chlorotic or necrotic lesions produced on transgenic leaves were only slightly lower (30–60%) than on non-transgenic plants. On the strength of all these results, it was concluded that the protection against virus infection blocks an early stage in the virus infection process which is not required for infection by viral RNA. Perhaps the protection operates by interfering with the uncoating of the virion, and/or preventing the translation of the RNA. For example, virus stripping may require association with a specific subcellular compartment or with a cell-specific function which is interfered with by the expression of the viral coat protein-coding sequence in transgenic plants.

In these experiments the TMV coat protein gene was derived from the common (U1) strain; however, protection was afforded against each of the strains of TMV so far tested, although protection was better against some challenge strains than against others. This was most notable in transgenic tomato plants that expressed the coat protein gene of TMV. When seedlings derived from transgenic tomato plants were infected with as much as 20 µg of U1 TMV per ml, none of the plants showed disease symptoms over the 30 day observation period. However, when inoculated with the more aggressive PV-230 strain, as many as 50% of the plants showed disease symptoms within 30 days, although all were delayed in disease development compared to non-transgenic plants. Protection is also afforded against other strains of TMV and tomato mosaic virus (L-TMV, ToMV-1, ToMV-2). The results of these experiments (R.S. Nelson et al, unpublished) indicate that protection by the expression of the coat protein gene operates against a large number of strains of TMV; however, the degree of protection may differ depending upon the strain. This is in contrast to the protection afforded by other genetic resistance loci, such as *Tm-1* and *Tm-2* loci in tomato (Nishiguchi & Motoyoshi, this volume). In these cases a high level of protection is afforded against most strains of ToMV, but little or no protection is afforded against other strains of ToMV (for example, *Tm-1* resistance is broken by the ToMV-1 strain of virus).

Similar experiments have been done with AlMV (Tumer et al 1987). Transgenic tobacco and tomato plants that express the AlMV coat protein-coding sequence accumulate the CP so that it constitutes approximately 0.5% of the cell protein. Seedlings derived from such transgenic plants were protected against inoculation by AlMV. As with the TMV example, the numbers of necrotic lesions on transgenic tomato plants infected with AlMV were lower than on control plants. Likewise, plants that expressed the CP gene of AlMV either did not become diseased or showed disease symptoms much later after inoculation than did the control plants. So far, the protection against TMV in plants that expressed the TMV CP gene has been mimicked in all respects by the protection against AlMV in transgenic plants that expressed that AlMV CP-coding sequence.

We concluded from these experiments with two different viruses, each having different architecture and mode of replication, that protection could be afforded by the expression of the viral CP gene. This leads to the suggestion that protection could be afforded against other viruses should a similar approach be taken.

The level of gene expression is correlated with the level of protection

There is a correlation between the level of gene expression of the coat protein-coding sequence in transgenic plants and the degree of protection. In our laboratory a chimeric gene was constructed with the transcriptional promoter of the 19S RNA of CaMV and the TMV CP gene. Although this promoter is known to cause the expression of genes in transgenic plants, its level of expression in solanaceous plants is markedly lower than that driven by the 35S promoter (Nagy et al 1985). We estimate expression from this promoter to be 20- to 50-fold lower than expression from the 35S promoter (M.A. Lawton & R.N. Beachy, unpublished). Although it was possible to detect the transcript of the 19S promoter:TMV CP-coding sequence in transgenic plants, the immunoassay was not sufficiently sensitive to detect TMV CP. These plants were not protected against infection by TMV (P. Powell Abel & R.N. Beachy, unpublished). A chimeric gene was also made with the promoter taken from the gene encoding the small subunit of *Petunia* ribulose bisphosphate carboxylase. Expression from this promoter caused lower levels of accumulation of CP (1/5 to 1/20 the level of expression produced by the 35S:CP gene construct). When these plants were infected with U1 TMV the level of protection was substantially less than in plants in which the 35S promoter was used (W.G. Clark et al, unpublished). These results indicate a direct positive correlation between the level of expression of the gene and the degree of protection. They also raise the possibility that increasing the level of expression of the coat protein-coding sequence may lead to increased protection.

Sequence specificity of the protection reaction

To determine whether protection could be afforded by the expression of sequences of TMV RNA other than the CP-coding sequence, chimeric genes which caused the expression of other TMV sequences were constructed. In these experiments the inverted orientation of nucleotides 3335 to 6395 of TMV RNA (the 3' half of the viral genome inverted so that the plant would accumulate antisense TMV RNA), and a gene encoding the 30 kDa protein of TMV (the protein responsible for cell-to-cell spread of TMV; Deom et al 1987) were expressed. Transgenic plants that expressed these sequences showed less than 1% of the protection against superinfection by TMV of those plants that expressed the TMV CP sequence (Beachy et al 1987b). If protection was afforded by expression of the other TMV sequences, the level was insufficient to allow detection under the assay conditions of the experiments. Evidently, expression of the CP-coding sequence is the preferred route for protection against TMV.

Protection in protoplasts of transgenic cells

We also demonstrated protection in protoplasts isolated from transgenic cells. It had been reported that 'classical' cross-protection could be demonstrated in protoplasts (Otsuki & Takebe 1976). We therefore isolated protoplasts from transgenic tobacco cells that expressed the coat protein gene and introduced TMV or TMV RNA by electroporation. In these experiments the number of transgenic protoplasts that became infected after poration in the presence of TMV was much lower (4–9%) than after infection of control protoplasts (40–60%). However, the yield of TMV per infected protoplast was essentially the same in both types of cells. As in whole leaves, inoculation of protoplasts with TMV RNA overcame the protection when high concentrations of TMV RNA were used. These results suggest that the protection mechanism is an intracellular rather than an extracellular process (J. Register & R.N. Beachy, unpublished results.).

Conclusion

The results of the experiments described here indicate that transgenic tobacco and tomato plants that express the coat protein-coding sequences of AlMV or TMV RNA exhibit a degree of protection against each of the respective viruses. Transgenic plants have many of the characteristics observed in some of the classical examples of cross-protection: that is, protection is greater against virions than against viral RNA; protection is expressed as a reduction in numbers of sites where infection take place; and protection is afforded against different strains of TMV (reviewed by Hamilton 1980). Unlike cross-protection, however, protection in transgenic plants is effective against distantly related strains of TMV, such as strains of tomato mosaic virus. In transgenic plants protection is effected by the expression of a single viral gene, not as a result of full virus infection or the resulting pathogenic states of the cell. Transgenic plants, therefore, offer a 'cleaner' system for the study of the protection reaction. If protection in some way interferes with the early stages in virus infection and replication, these plants provide an excellent system in which to study those early events. Likewise, if there are multiple mechanisms which together result in a protected phenotype, transgenic plants again provide the medium for the study of these mechanisms.

Lastly, transgenic plants and their progeny are protected from virus infection in the same way as expected for any dominant Mendelian trait. We have noted no differences in plant growth and development or in fruit-set of transgenic plants compared to non-transgenic plants. We therefore propose that expression of such genes may provide protection without a substantial loss in productivity of the transgenic plant, as can occur with classically cross-protected plants. This approach also offers an alternative to the plant breeder who searches for resistance genes. It often takes many years for the incorporation of a resistance gene into an agronomically important genetic background. The protection afforded by genetic transformation, although limited to a relatively small number of plant species at present (for technical reasons), can potentially be applied to a wide variety of viruses. The transformation approach can thus be seen as a supplementary approach to resistance against virus diseases that can be used by plant breeders.

References

Beachy RN, Rogers SG, Fraley RT 1987a Genetic transformation to confer resistance to plant virus diseases. In: Setlow J (ed) Genetic engineering. Plenum Press, New York, vol 9:229–247

Beachy RN, Stark DM, Deom CM, Oliver MJ, Fraley RT 1987b Expression of sequences of tobacco mosaic virus in transgenic plants and their role in disease resistance. In: Bruening GB et al (eds) Tailoring genes for crop improvement: an agricultural perspective. Plenum Press, New York, in press

Deom CM, Oliver M, Beachy RN 1987 The 30-kilodalton gene product of tobacco mosaic virus potentiates virus movement. Science (Wash DC) 237:389–394

Fraley RT, Rogers SG, Horsch RB 1986 Genetic transformation in higher plants. CRC Crit Rev Plant Sci 4:1–46

Fraley RT, Rogers SG, Horsch RB et al 1985 The SEV system: a new disarmed Ti plasmid vector system for plant transformation. Bio-Technol 3:629–635

Goelet P, Lomonosoff GP, Butler PJG, Akam ME, Gait MJ, Karn J 1982 Nucleotide sequence of tobacco mosaic virus. Proc Natl Acad Sci USA 79:5818–5822

Hamilton RI 1980 Defenses triggered by previous invaders: viruses. In: Horsfall JG, Cowling EB (eds) Plant disease: an advanced treatise. Academic Press, New York, vol 5:279–303

Horsch RB, Fry JE, Hoffmann NL, Eickholtz D, Rogers SG, Fraley RT 1985 A simple and general method for transferring genes into plants. Science (Wash DC) 227:1229–1231

Loesch-Fries LS, Jarvis NP, Krahn KJ, Nelson SE, Hall TC 1985 Expression of alfalfa mosaic virus RNA 4 cDNA transcripts in vitro and in vivo. Virology 146:177–187

Maniatis T, Fritsch EF, Sambrook J 1982 Molecular cloning: a laboratory manual.

Cold Spring Harbor Laboratory, Cold Spring Harbor, New York.

Nagy F, Odell JT, Morelli G, Chua N-H 1985 Properties of expression of the 35S promoter from CaMV in transgenic tobacco plants. In: Zaitlin M et al (eds) Biotechnology in plant science: relevance to agriculture in the eighties. Academic Press, New York, p 227–235

Nishiguchi M, Motoyoshi F 1987 Resistance mechanisms of tobacco mosaic virus strains in tomato and tobacco. In: Plant resistance to viruses. Wiley, Chichester (Ciba Found Symp 133) p 38–56

Nelson RS, Powell Abel P, Beachy RN 1987 Lesions and virus accumulation in inoculated transgenic tobacco plants expressing the coat protein gene of tobacco mosaic virus. Virology 158:126–132

Otsuki Y, Takebe I 1976 Double infection of isolated tobacco leaf protoplasts by two strains of tobacco mosaic virus. In: Tomiyama K et al (eds) Biochemistry and cytology of plant–parasite interaction. Kodansha, Tokyo, p 213–222

Powell Abel P, Nelson RS, De B et al 1986 Delay of disease development in transgenic plants that express the tobacco mosaic virus coat protein gene. Science (Wash DC) 232:738–743

Rogers SG, O'Connell K, Horsch RB, Fraley RT 1985 Investigation of factors involved in foreign protein expression in transformed plants. In: Zaitlin M (ed) Biotechnology in plant science: relevance to agriculture in the eighties. Academic Press, New York, p 219–226

Rogers SG, Horsch RS, Fraley RT 1986 Gene transfer in plants: production of transformed plants using Ti plasmid vectors. Methods Enzymol 118:627–640

Sherwood JL 1987 Mechanisms of cross-protection between plant virus strains. In: Plant resistance to viruses. Wiley, Chichester (Ciba Found Symp 133) p 136–150

Tumer NE, O'Connell KM, Nelson RS et al 1987 Expression of alfalfa mosaic virus coat protein gene confers cross-protection in transgenic tobacco and tomato plants. EMBO (Eur Mol Biol Organ) J 6:1181–1188

DISCUSSION

Matthews: Dr Beachy, in a tomato glasshouse the concentration of inoculum involved in accidental mechanical transfer of virus might be approximately 1 mg per ml spread over a small area, perhaps about 0.5 cm^2 of leaf. Have you considered testing your transgenic plants by using such a concentrated inoculum on a small area?

Beachy: We have been trying to find out how much inoculum might be involved in natural transmission. We have been using 20 μg per ml of inoculum over the total area of two leaves for these tomato plants. We should perhaps apply a more concentrated inoculum at one site and repeat the tests.

Matthews: What about the possibility of successive transmissions of your virus from the first transgenic plants that became systemically infected? Might successive transfers select an altered virus to which the transgenic plants are not resistant?

Beachy: We have protection against many TMV strains and we would like to find one that is not protected against, so we are doing that experiment. We propagate the virus at a relatively high level, in order to get systemic infection, and then passage it. We have done seven cycles and disease development is no more rapid in the latest cycles than in the original infection. We shall do 20, 30 and 40 cycles, and then do experiments with specific infectivity reactions. We would like to isolate a strain that overcomes the protection, because it would give us a clue as to what is required for protection.

van Kammen: You used the aggressive strain of TMV, PV-230, which overcomes protection at a lower inoculum concentration than the U1 strain, against which maximal protection was obtained. Did you compare the genetic differences between strain PV-230 and the normal strain? In particular, did you look for virus property(ies) which make a strain more, or less, aggressive? That might provide some information about how the protection works.

Beachy: We have cloned cDNAs of this strain but have not characterized them yet. We want to exchange genomic sequences between the virus strains and make recombinants.

Zaitlin: We recently collaborated with Dr Graham Hills at the John Innes Institute. We investigated a large number of tobacco cells by immunostaining for the 126K protein. In no case did we observe more than one pocket per cell of the viroplasm which contains this 126K protein. It is possible that cyclosis has caused aggregation and that initially there were a number of sites.

Beachy: We favour the hypothesis that a membrane- associated component is involved in the protection because of the membrane association of the replication complex. I don't know how many sites there are for virus replication in a cell, but there may be a limited number of them.

Fritig: You have shown a reduced number of possible infection sites in the local lesion host. What is the effect on lesion size?

Beachy: The lesion sizes and the rates of growth of the lesions were the same in the transgenic and control plants.

Fritig: Does this prove that the constitutive expression of the coat protein in the transgenic plants does not interfere with virus replication?

Beachy: Our work on inoculation with TMV RNA leads us to that conclusion and we are close to proving it.

Hohn: During virus infections, coat protein can enter chloroplasts and cause damage there. I wonder whether you observed any irregularities in your transgenic plants caused by accumulation of coat protein in the chloroplasts?

Beachy: In virus-infected cells about 0.1% of the total coat protein is inside chloroplasts. We also find about 0.1% of the coat protein in transgenic cells inside chloroplasts, but the level is much less than in virus infection. We do find the coat protein in both the stroma and thylakoids in these chloroplasts. Therefore it is not necessary to have a disease state for the virus coat protein to enter chloroplasts. We did a glasshouse yield experiment on transgenic and control plants and found no difference in growth rates, numbers of flowers, and the fresh and dry weights of different parts of the plant. In fact, after 60 days, disease symptoms developed in the glasshouse and almost all of the control plants became infected, whereas none of the transgenic plants were affected.

Goldbach: Dr Gus De Zoeten tested your transgenic plants and found them to be more sensitive to dryness than the controls. They seem, in general, to be more stress sensitive.

Beachy: We have studied eight different tobacco lines and three different tomato lines and we don't see any great differences in plant growth. One tobacco line has a smaller root mass. Perhaps he was working on that line. One of the eight tobacco lines was abnormal in that it had slightly elongated leaves. The others appeared completely normal. When studying transgenic plants, one should always ensure that the plants are normal throughout. One might expect minor changes in some plants due to somaclonal effects. We have not noticed differences in properties such as sensitivity to virus infection.

White: Have you inoculated your transgenic plants with any completely unrelated viruses?

Beachy: Yes, we have used cucumber mosaic virus, alfalfa mosaic virus, potato viruses X and Y, and tobacco etch virus. We inoculated transgenic plants which expressed the TMV coat protein gene, plants with non-expressive cell lines and normal Xanthi plants. We scored the rate of disease development and the amount of virus accumulation in inoculated leaves and in leaves above inoculated leaves.

When a low level of PVX inoculum was used there was an equal amount of virus accumulation in the inoculated leaves but reduced systemic spread in transgenic plants that expressed the coat protein. When we increased the level of inoculum fivefold, there was no protection.

When we repeated the experiments with CMV we initially concluded that there was no protection but we subsequently discovered, by constructing an inoculum dilution curve, that we had inoculated at a supersaturating level. Using just sufficient levels of inoculum to ensure that all the controls became diseased, we observed protection in the transgenic plants. As with PVX, when the inoculum concentration was increased fivefold there was no protection. The same was true with the other viruses tested: protection was seen at a low level of inoculum but lost at higher levels. I suspect that this resistance against other viruses is due to a second mechanism that may be unrelated to the mechanism that provides resistance against TMV.

Zaitlin: Presumably this has nothing to do with the relative infectivity of RNA and virus?

Beachy: I don't know. We have only tried virions of these other viruses, not viral RNAs.

Davies: Tobacco rattle virus might be interesting to try, because there is no homology with the coat protein but high homology with the 30K protein. Also it would be possible to experiment with the RNA-1 alone.

Beachy: We have used tobacco rattle virus: we saw a reduction in numbers of necrotic spots but we still found systemic disease.

Davies: Did you study the RNA-1 component alone?

Beachy: No; that would certainly be a good experiment. We would need good antibodies and good detection methods.

Harrison: What strain of tobacco rattle virus did you use?

Beachy: The CAM strain that we obtained from Dr David Bisaro.

Davies: In fact that has now been designated pepper ringspot virus (Robinson & Harrison 1985).

van Vloten-Doting: Did you obtain the same results with PVY, which belongs to the other superfamily?

Beachy: All the viruses tested, including PVY, caused slow disease development in the transgenic plants. Some of these experiments were done by immuno dot blots, which involve a great deal of work, so others have been done simply by monitoring disease development. We have done dot blots for PVX and AlMV and the results demonstrate that the movement protection can be overcome. For PVY and tobacco etch virus, we studied only disease development, which is delayed. As many as 50 or 60% of the plants never showed disease symptoms at a low level of inoculum. This indicates another broad protection mechanism.

Bol: What copy number of the coat protein gene do you find in the plants?

Beachy: In 80% of the plants there is a single copy of the gene.

Bol: Do you see a segregation ratio of 3:1 for the coat protein gene in your next generation of plants?

Beachy: There is a good correlation between the segregation ratio and protection. We have occasionally come upon a plant that we had scored as negative, as not expressing coat protein, but it showed resistance. However, when we checked it we found that it was expressing nopaline and thus was a transgenic plant.

Zimmern: Do you see any coat protein outside the cells?

Beachy: We looked outside the cells by preparing cell wall fractions using the same method as Dr Joe Varner (Washington University). We washed them extensively with calcium chloride and still found some coat protein associated, but we cannot prove that this was not due to unbroken cells. We think there is a small amount of coat protein in cell walls. Dr Gus De Zoeten took our transgenic plants to look for coat protein in cell wall washings, but found none, to the best of my knowledge. Perhaps there is too little coat protein made in these cells relative to virus infection for substantial amounts of coat protein to be outside the plasma membrane. I do not think these observations contradict previous conclusions.

Harrison: What happens when you inoculate the transformed plants with, for example, the RNA of strains PM2 or DT1-G of TMV?

Beachy: We have tried PM2 but we did not have much success at inoculating RNA. We were not able to get infection.

Harrison: We have done that experiment. With DT1-G, which is the more suitable strain because it produces no coat protein at all, we could not detect virus particles in inoculated leaves of tobacco plants transformed with the TMV coat protein gene.

Beachy: You could not detect encapsidation. Did you have a level of coat protein that was comparable to that in our plants?

Harrison: I don't know. It's hard to know whether there was enough coat protein there or not.

Beachy: That is why we are preparing the cross between the 30 kDa plant and the coat protein plant.

Harrison: The problem might be that the cells do not contain the aggregates of TMV coat protein that are needed to initiate assembly of TMV nucleoprotein particles.

Beachy: That is probably right, because aggregation and state of assembly are concentration dependent. We may just not have enough there to make a 20S disk of the aggregated TMV-capsid protein.

Dodds: Are you doing parallel experiments, infecting non-transformed plants with U1 TMV and then doing a classical cross-protection experiment with the low levels of challenge inoculum that you find to be important in experiments with transgenic plants? For example, what happens if you challenge-inoculate TMV-infected plants with small inocula of the four or five unrelated viruses that you have studied?

Beachy: We have not done those experiments. I didn't expect them to work. The results would be difficult to interpret. When attempts are made to infect a diseased plant at a low dosage of challenge inoculum, it may not be easy to distinguish between a change in physiological status of the plant that prevents infection and protection against the virus, which would be minimal compared to protection against TMV.

Zaitlin: We don't know much about symptom determinants. Are there constructs in which portions of the viral genome can be inserted in the plant, resulting in mimicry of virus symptoms?

Beachy: That has been demonstrated only in cauliflower mosaic virus (CaMV). We now have transgenic plants with the 30K gene and others with the coat protein and we observe no disease development in either type. We conclude that either these gene products are not involved in symptom development, or there is too low an expression level to cause symptoms.

Bruening: Dr R.J. Shepherd's group has transformed tobaccos with gene VI of CaMV. The plants had mosaic symptoms and looked as if they were infected.

Davies: Do these transgenic plants have inclusion bodies?

Bruening: They have gene VI protein, but no inclusion bodies.

Beachy: In several of Dr Shepherd's tobacco plants of increasing age, the lower leaves had the greatest amount of chlorosis and the upper leaves were vein-cleared only. As the plants aged they became fully chlorotic—no green islands remained.

van Vloten-Doting: Is there any difference in expression in the cells showing the mosaic or are they all expressing the same amount of protein? We have already mentioned that dark green parts of leaves contain less virus than the other parts. Working with the geminiviruses, Rogers et al (1985) made the cross between the transgenics containing the A and B parts of the genome and obtained mosaic symptoms. One might expect every cell to contain the same amount of virus. Nevertheless they observed the typical mosaic symptoms.

Beachy: Presumably, not every leaf is expressing the geminivirus genome, because DNA excision events affecting both components must occur in the same cell in order to produce symptoms.

van Vloten-Doting: Is there a correlation between the excision and the production of virus and symptoms?

Beachy: Those experiments have not been done.

van Vloten-Doting: Genetically all those cells should be identical.

Beachy: Yes, but the excision of DNA would probably be a relatively rare event, and then replication is limited to cells which are sufficiently connected to give spread of the disease.

van Vloten-Doting: Only one viral protein is produced in transgenic plants expressing the CaMV gene VI. In principle this protein is produced in every cell, so why do symptoms develop?

Beachy: Symptoms become more severe as the gene product accumulates. The older cells accumulate more protein and the younger cells have less, and therefore less disease.

van Vloten-Doting: In that case the whole leaf should show the same amount of chlorosis.

Beachy: It does show very uniform chlorosis.

Harrison: Dr Beachy, is there an advantage in increasing the number of insertions of a viral coat protein gene into the DNA of plants? Does one see increasing gene expression with an increased number of insertions?

Beachy: We tried to correlate the numbers of genes inserted, the amount of RNA transcribed and the amount of protein produced. There is a correlation between the amount of RNA and the amount of protein, but the numbers of genes and the amount of RNA are not highly correlated. Other people have made similar observations in other transgenic plants. Presumably these multiple insertion events sometimes result in inactivated genes and rearrangements.

If we have a plant with two or three copies of the gene where there is no rearrangement, we get slightly more RNA. We have not been able to correlate the increase in amount of RNA with an increase in resistance. I suspect it is because our biological assays are not sensitive enough. We shall have to obtain at least a five- to tenfold increase in gene expression before we make those correlations.

Harrison: If you have two effective transformations which may only involve single insertions, is there an advantage in crossing those plants?

Beachy: We have taken these plants to homozygosity and crossed them back to the parents, and we shall have a whole population of plants that will have one gene expressed, a population that will have two genes expressed, and a population with none. Then we can ask if we are able to quantify the resistance. Is there a gene-dosage effect? I would expect so but we cannot answer that question yet.

Harrison: One might expect to get variable levels of expression from random insertions.

Beachy: That is an interesting point. We have not observed that. In six different plants, each of which had single genes inserted, we got a variation of about two- or threefold of the amount of RNA and the amount of protein. We did not find the tenfold variation in amount of protein that has been reported with other transgenic traits. With the small subunit promoter we found a tremendous variation. If there is a single copy of a gene in one plant and a single copy of a gene in a second plant, they will have very different amounts of RNA with the promoter that we use. This may reflect on our construct: the relationship of the promoter to the 5' end, the relationship of the 5' end to the structural region of the gene, and so on. The work of C. Dean and P. Dunsmuir, in California, would suggest that RNA structural features are important to obtain efficient and reproducible gene expression.

Bol: We have transformed Samsun NN tobacco plants with DNA copies of alfalfa mosaic virus (AlMV) genomic RNAs 1, 2 and 3, and subgenomic RNA-4. In non-transformed plants, the genomic RNAs are not infectious unless one adds RNA-4 or coat protein. There is a specific binding site for coat protein at the 3' end of the viral RNAs. Plants were also transformed with a DNA construct containing 16 tandem copies of this coat protein binding site. Transgenic plants were obtained expressing plus-strand and minus-strand AlMV RNAs 1, 2, 3 and 4 from nuclear genes. On average the minus strands accumulated to a lower level than the plus strands. In contrast to Dr Beachy's result, we saw no processing of plus-strand RNA-4. However, we observed processing of plus-strand RNA- 3. Plants with a full-length copy of RNA-3 inserted in their genome produced a transcript of subgenomic size, containing sequences of the 32K gene fused to the 3'terminal sequence of RNA-3. This suggests a splicing event.

Plants producing AlMV RNA-4, 'CP plants', accumulated viral coat protein to a level of about 0.05% of the total soluble protein. These plants were tested for cross-protection. As controls we used non-transformed Samsun NN tobacco and tobacco transformed with the vector alone without viral cDNA. At a concentration of 2.5 µg/ml, the AlMV strain YSMV induced approximately 500 lesions per half leaf in the controls but none in the CP plants. When leaf homogenates of these latter plants were inoculated onto control plants, no lesions developed, indicating absence of cryptic infection in the transgenic plants.

CP plants and controls were equally well infected with a mixture of AlMV

RNAs 1–4, showing that protection is not active against isolated RNA. If inoculated with RNAs 1, 2 and 3, only the CP plants became infected but not the controls. Apparently the endogenously produced coat protein is able to replace the coat protein in the inoculum. Probably, this coat protein is binding to the 3' termini of the incoming RNAs, thus permitting the replication of these RNAs. Apparently, the parental RNAs are not fully encapsidated by the coat protein produced in the CP plants (Van Dun et al 1987).

We have also transformed plants with the coat protein gene of a strain of tobacco rattle virus. These plants, and those expressing AlMV non-structural genes, have not yet been tested for cross-protection.

Beachy: In their experiment with AlMV coat protein expression, Tumer et al (1987) found the level of protein accumulation to be about 0.5%, whereas you reported about 0.05% in yours, Dr Bol. I don't know what the difference is.

Bol: The inaccuracy in estimating expression levels from Western blots may be responsible for part of the difference.

Davies: Have you looked for 'stripposome' activity in epidermal cells directly, using the technique of Shaw et al (1986)?

Beachy: The first approach will be to make high pH-treated rods and see if there is protection against them also in these plants.

Davies: Yes, but they have detected the stripposomes *in vivo* in epidermal cells. To use ^{35}S-labelled or ^{3}H-labelled viruses in protoplasts or leaves is the most obvious way.

Beachy: We would like to do that experiment.

Davies: Would it be useful to make a construct with the omega sequence attached to your coat protein gene to see if it makes a difference to cross-protection?

Beachy: The omega fragment is derived from the 5' end of TMV RNA (produced by treating the RNA with RNAse T_1) and includes the untranslated region of the TMV RNA molecule. When Mike Wilson at the John Innes Institute put that fragment in front of the CAT mRNA he obtained increased translation of the mRNA. This indicates that the omega fraction is probably a good leader for translation. The increase was between three- and tenfold *in vivo* as well as *in vitro*.

Jobling & Gehrke (1987) demonstrated that the 5' untranslated region of RNA-4 of AlMV can stimulate a 35-fold increase in *in vitro* translation of heterologous mRNA. This is an alternative sequence for increasing the translational efficiency of a molecule.

Davies: It provides a phenotypic way of making an increased gene dosage.

Beachy: Correct. We have also changed the 3' end of the coat protein mRNA by removing the tRNA-like structure and attaching the 3' end of another mRNA that we have been working with. By altering sequences at both the 5' and 3' ends we hope to increase the level of protein accumulation without changing the level of promoter expression.

Davies: Have you looked to see if there is a host-induced RNA–RNA polymerase in the transgenic plants?

Beachy: No, we have not.

White: Might your transgenic plants be showing resistance to mechanical inoculation?

Beachy: I doubt it, because our transgenic control plants do not show protection. They express kanamycin and nopaline but not the coat protein gene. We also have plants that express other genes from the 35S promoter, such as the CAT gene and the seed storage protein gene. The CAT gene is enzymic and does not accumulate, but the storage protein accumulates to low levels in leaves: these plants need to be tested for protection, and may be the best controls for addressing the question you raise.

White: It would be interesting to compare the resistance of your transgenic plants to both aphid and mechanical inoculation of the same virus.

Beachy: We hope to do those experiments.

Goldbach: Dr Bol, with AlMV RNA-4 you saw cross-protection. What happened to the plants transformed with copies of RNAs 1, 2 and 3?

Bol: We have not got those results yet.

Goldbach: Do you produce any damage of the plant phenotype in other respects by introducing viral genes in tobacco?

Bol: We see no difference between the transformed and untransformed plants.

Davies: That implies that no gene is individually responsible for symptoms.

van Vloten-Doting: No single gene is responsible for symptoms at the level at which it is expressed: that is an important point.

Hohn: If man can achieve resistance by these transformations, perhaps nature has already done this? Are there plant genomes which are virus resistant because part of a virus genome has been incorporated?

Beachy: For instance, are there any TMV sequences already incorporated in the tobacco genome and are those being expressed and responsible for an existing resistance trait?

Zaitlin: We recently reported cross-reactivity between the large subunit (Dietzgen & Zaitlin 1987) of the host protein ribulose-1,5- bisphosphate carboxylase and TMV coat protein at the protein level, but not at the nucleic acid level.

Beachy: We made a riboprobe of the TMV coat protein gene to look at the organization of the coat protein gene in our transgenic plants. We only found DNA fragments of the expected size. We don't find other sequences that would be reminiscent of the insertion of TMV coat protein sequences.

Sela: There are plenty of genomic libraries now—this possibility could be screened out.

Beachy: I don't think it is worth doing.

Rybicki: Given that genetic recombination between RNA components of plant viruses can occur (Bujarski & Kaesberg 1986), is it possible that super-infection of, for example, transgenic plants expressing TMV coat protein mRNA, with another strain of the virus, could result in recombination events leading to the emergence of recombinant strains?

Beachy: A system for achieving that would be useful. However, the model of recombination between RNA molecules evokes the involvement of double-stranded RNA, and the juxtaposition of replication complexes would be necessary to allow strand displacement and subsequent exchange. A high level of RNA polymerase or replicase would be needed to detect such events. In addition, the 5' end of TMV RNA is not present in our construct, which presumably would preclude the replicase binding to it during any replication. Therefore the complement would not be achieved.

Harrison: If viruses replicating in plants transformed with a viral coat protein gene could acquire this gene by RNA recombination, this would be important for vector-borne viruses in which the coat protein plays a key role in transmission. Any such phenomenon would be of interest to the authorities who control release of genetically engineered products.

van Vloten-Doting: Not only recombination but also transencapsidation may be very important.

Bol: What are the prospects for introducing transgenic plants in the field in the United States?

Beachy: We have submitted an application for release but we do not know yet what the response will be. The important issues, such as the release of germ plasm that might spread to weed species, and potential problems with selecting out other highly virulent strains, have yet to be resolved. We hope that permission will be granted, but the questions of control must be addressed. We now have to face problems that plant pathologists have had to face for some time. It appears that we may need permission not only to put out transgenic plants but also to make inoculations in the field.

Sänger: Have transgenic plants already been introduced in the field?

Beachy: There have been three examples of release of transgenic tobacco plants in the field in the USA. We hope to be granted permission to put the plants out and to test them with virus. In addition, we would like to go to fruit: in tobacco the top of the plant is cut off but in tomato, fruit is produced. We have to design an application that would be compatible with these aims. This presents new questions to the regulatory agencies.

Harrison: Are you more likely to be allowed to work on tomatoes grown in glasshouses, where you could have a containment arrangement?

Beachy: Yes, under existing guidelines a well-controlled glasshouse with limited access is a permitted environment for these experiments.

van Kammen: Dr Beachy, you used the same method as Dr Fritig for detecting infection centres and you found a decreased number of successful infections, presumably because the sensitivity of the sites has decreased. In connection with Dr Fritig's paper we discussed stress as a cause of decreased sensitivity, although that occurred only in a limited zone of tissue. Is the protection in your plants comparable to the decreased sensitivity induced by stress?

Beachy: I think it is different: we were able to establish infection with viral RNA very easily, but we couldn't establish infection with the micro-inoculation technique using virus rods, even though we used TMV at 500 μg per ml as the inoculation.

van Kammen: The question is what triggers the stress reaction in a virus-infected plant. Your protected plants produce viral coat protein. What more, if anything, is required for a stress reaction that results in decreased sensitivity to occur?

Beachy: I can't address that question, because I don't know what to measure.

Hohn: What about transgenic plants with other TMV proteins?

Beachy: The only other protein we have expressed is the 30K gene. We can prove the 30kDa gene product is there, but it gives no protection against subsequent infection. No one has yet checked the 126K gene, to the best of my knowledge.

Zaitlin: Dr Bol, is there a problem with the fact that you are putting your sequences into the nucleus and then hoping they would be expressed elsewhere?

Bol: RNA-3 was found to be processed in the nucleus, but the others are not processed. With RNA-4 there is an efficient translation.

Zaitlin: Protection may only occur if those sequences are in a site in the cell where they would be in contact with the incoming virus.

Bol: The observation that the RNAs 1, 2 and 3 are able to infect those plants shows that interaction of the coat protein with the incoming RNA is possible.

Zaitlin: With the coat protein, yes, but interacting with the RNAs 1 and 2 themselves? Somehow, as in Dr Beachy's results, the coat protein that accomplishes this is everywhere in the cell!

Beachy: The ability of some viral capsid proteins to cross membranes is interesting. Is there some hydrophobicity or other feature of the capsid protein that facilitates its movement through membranes? In the TMV case this includes the chloroplast, and the protein presumably enters as subunits. What about the uptake of a virus rod through the plasma membrane?

Bol: The alfalfa mosaic virus RNA-4, in addition to being a messenger for the coat protein, also has a high affinity binding site for the coat protein which may influence its translation. We have not compared the effect of TMV coat protein gene with or without the origin of assembly.

Beachy: In the U1 strain the origin of assembly is in the 30K gene sequence, not on the coat protein gene itself.

Bol: Have you used coat protein genes containing the origin of assembly?

Beachy: No. This will be done in other plants which contain the 30K plus the coat protein.

Harrison: One could do that with the sunn hemp strain of TMV, which has the origin of assembly sequence in the coat protein gene.

Bol: We have not seen viral particles in the plants producing coat proteins; perhaps the RNA-4 is encapsidated by coat proteins.

Nishiguchi: The extent of cross-protection with the PV-230 severe strain of TMV is less than with the other strains. Why is this?

Beachy: We don't know. PV-230 is not a well-characterized strain, so we cannot generalize.

Nishiguchi: How does the multiplication of PV-230 compare with that of other strains?

Beachy: If one infects with U1 or PV-230, as well as a mild strain of TMV, the amount of virus accumulation per cell is essentially equivalent.

Nishiguchi: How does the amino acid sequence of the PV-230 coat protein compare with the sequences of coat proteins of other TMV strains?

Beachy: We don't know the answer to that question yet.

References

Bujarski JJ, Kaesberg P 1986 Genetic recombination between RNA components of a multipartite plant virus. Nature (Lond) 321:528–531

Dietzgen RG, Zaitlin M 1987 Tobacco mosaic virus coat protein and the large subunit of the host protein ribulose-1,5-bisphosphate carboxylase share a common antigenic determinant. Virology 155:262–266

Jobling SA, Gehrke L 1987 Enhanced translation of chimaeric messenger RNAs containing a plant viral untranslated leader sequence. Nature (Lond) 325:622–625

Robinson DJ, Harrison BD 1985 Unequal variation in the two genome parts of tobraviruses and evidence for the existence of three separate viruses. J Gen Virol 66:171–176

Rogers S, Horsch R, Hoffmann N, Brand L, Cheng I, Suner G, Bisaro D 1985 Stable integration of geminivirus genomes into *Petunia* plants (abstr). First Int Congr Plant Mol Biol, Savannah, Georgia

Shaw JG, Plaskitt KA, Wilson TMA 1986 Evidence that tobacco mosaic virus particles disassemble cotranslationally *in vivo*. Virology 148:326–336

Tumer NE, O'Connell KM, Nelson RS et al 1987 Expression of alfalfa mosaic virus coat protein gene confers cross-protection in transgenic tobacco and tomato plants. EMBO·(Eur Mol Biol Organ) J 6:1181–1188

Van Dun CMP, Bol JF, van Vloten-Doting L 1987 Expression of alfalfa mosaic virus and tobacco rattle virus coat protein genes in transgenic tobacco plants. Virology 159:299–305

Resistance to viral disease through expression of viral genetic material from the plant genome

D.C. Baulcombe, W.D.O. Hamilton, M.A. Mayo* and B.D. Harrison*

*Plant Breeding Institute, Maris Lane, Trumpington, Cambridge CB2 2LQ and *Scottish Crop Research Institute, Invergowrie, Dundee DD2 5DA, UK*

Abstract. It has been predicted that expression of viral sequences in transformed plants may result in resistance to viral infection. This paper describes and evaluates examples in which this prediction has been tested. When the viral sequence used was based on the satellite RNA of cucumber mosaic virus (CMV) the approach was successful and the transformed tobacco plants did not show the typical CMV symptoms in the systemically infected leaves. This attenuation was not necessarily a result of an inhibitory influence of satellite RNA on viral RNA replication and so is thought to involve a direct interference between satellite RNA and the symptom-inducing capability of the virus. Other viral sequences have also been expressed in transformed plants in attempts to produce virus-resistant or tolerant plants. The degree of success has varied. In experiments in which different regions of the tobacco rattle virus (TRV) genome were expressed as antisense RNA in tobacco, the transformed plants did not show resistance to infection by TRV, possibly because the encapsidated viral RNA is not accessible to the antisense RNA molecules. Expression of viral coat protein in transformed plants did produce resistance to virus infection, but the resultant protection had different properties from the resistance produced by expression of satellite RNA sequences.

1987 Plant resistance to viruses. Wiley, Chichester (Ciba Foundation Symposium 133) p 170–184

The plant genes which affect susceptibility to viral infection have so far proved inaccessible to molecular analysis, although progress in this direction is illustrated by the discussion in this symposium of various mechanisms by which plants resist viral infection. This difficulty has stimulated a number of research groups to explore the use of genes from non-plant sources in transformation experiments designed to engineer virus-resistant plants. Beachy et al (this volume) have given an example of this approach in which the expression of viral coat protein genes in transformed plants mimicked the phenomenon of cross-protection between virus strains. In this paper we shall describe and evaluate two other types of experiments. These are similar to the work described by Beachy (Powell Abel et al 1986) in that the sequences expressed in plants are viral in origin. However, our experiments focus on different types of antiviral mechanisms which do not involve expression of viral protein in the transformed plants.

Satellite RNA of cucumber mosaic virus

Our experiments with cucumber mosaic virus (CMV) satellite RNA were originally designed to mimic a defective interfering (DI) system such as is found in animal viruses. CMV satellite RNA is a short RNA species found associated with some, but not all, natural isolates of CMV (reviewed by Kaper & Waterworth 1981). It is not obviously homologous with viral genomic RNA (Lot et al 1977) and therefore cannot be considered strictly as a DI RNA. However, satellite RNA has effects similar to those of DI RNA molecules in that it inhibits the replication of the viral RNA, probably by competing with it for replicase enzymes (Piazzolla et al 1982). There is also a reduced amount of viral RNA in systemically infected tobacco leaves when satellite RNA is included in the viral inoculum (Mossop & Francki 1979). This reduction may be related to the competitive effect.

CMV satellite RNA molecules also cause a modification of symptom expression by the virus. With certain types of satellite RNA, which we refer to as benign satellites, the effect is to suppress, more or less completely, the production of symptoms by CMV so that infected plants are not easily distinguishable from healthy plants (Jacquemond & Lot 1981). This phenomenon of satellite-mediated protection has been exploited agronomically in China with pepper plants (Tien Po et al 1987). After a prophylactic infection with CMV inocula containing a benign satellite there was an increase in the yield of fruit of 11–56%, depending on the geographical site. Presumably this effect derives from protection against the effects of adventitious infections by naturally occurring CMV.

The aim of our experiments was to introduce a DNA sequence coding for satellite RNA into the genome of transformed plants so that satellite RNA would be produced constitutively in those plants. It was predicted that, if this transcribed satellite RNA had the biological properties of naturally occurring satellite RNA, the transformed plants would show attenuated symptoms when inoculated with CMV. It was also predicted that less virus would accumulate in the systemically infected leaves than in plants that were not producing satellite RNA.

The vector molecule used for transfer of new DNA into the plant genomes was based on the tumour-inducing (Ti) plasmid of *Agrobacterium tumefaciens*. The oncogenic capability of the Ti DNA been deleted, so that morphologically normal and fertile plants could be regenerated from the transformed cells. Further important features of the promoter sequence and

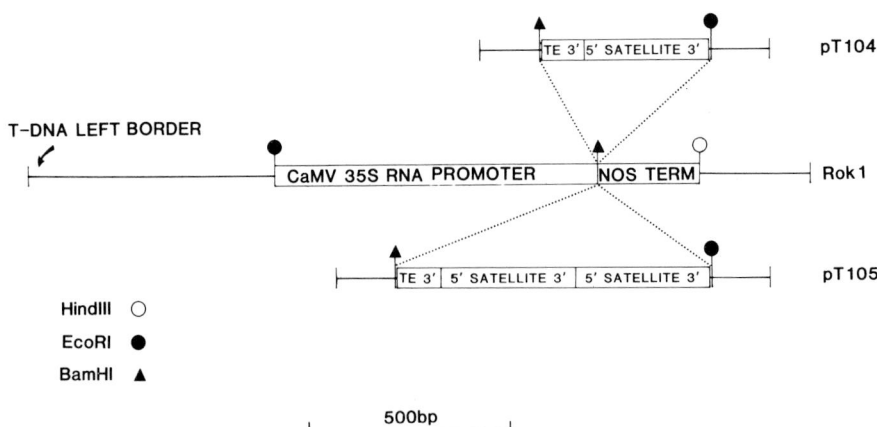

FIG. 1. DNA coding for satellite RNA structures introduced into transformed plants. Multimeric forms of complementary DNA of CMV satellite RNA were transferred from plasmids pT104 or pT105 and introduced into the expression cassette of the Ti plasmid-based vector, *Rok*1. The diagram shows incomplete (TE 3') and complete (5' SATELLITE 3') units of CMV satellite cDNA, the promoter sequence of the expression cassette (CaMV 35S promoter) and the transcriptional terminator sequence (nopaline synthase, NOS TERM) in the vector.

the form of the satellite sequences are illustrated in Fig. 1. The promoter was from cauliflower mosaic virus (CaMV), for the 35S RNA, and is known to be strong and constitutively active. The satellite sequences were derived from two complementary DNA (cDNA) constructions, in plasmids designated pT104 and pT105. These plasmids contained multimeric satellite cDNA comprising 1.3 (pT104) or 2.3 (pT105) units. It was thought that transcripts of these multimers would mimic possible replication intermediates of CMV satellite RNA. The likely advantage of this would be that any adverse effects of terminal sequences derived from the promoter or transcriptional terminator on the biological activity of the satellite transcripts would be confined to outer units of the multimer.

In the transformed tobacco plants produced using these Ti plasmid vectors, and in the absence of CMV, the amount of the multimeric satellite transcripts was typical of other RNA molecules transcribed by the CaMV 35S RNA promoter (Baulcombe et al 1986). The unit-length satellite RNA, which is the characteristic form of satellite RNA in natural infections, was not detected. After infection of the transformed plants with a satellite-free isolate of CMV the satellite RNA accumulated at much higher levels than before, indicating that the virus could replicate the transcribed, multimeric form of the satellite RNA. It was also noted that the satellite produced after CMV infection was primarily of unit length and there was as much of it as if the satellite RNA had been introduced by inoculation (Baulcombe et al 1986). These findings suggested that the transcribed satellite was replicated by the processes typical for natural satellite RNA. We were therefore not surprised to find that the presence of the transcribed satellite resulted in the predicted effects on the replication of CMV RNA and symptom production.

Thus, infection of transformed (satellite-producing) plants with CMV led to a reduction in the amount of viral RNA in systemically infected tobacco leaves. The final amount was 20% or less of that found in plants not producing satellite RNA. Furthermore, the satellite-producing plants did not develop the mosaic which is typical of CMV (Harrison et al 1987) and appeared similar to healthy tobacco plants.

Surprisingly, however, the inhibition of viral RNA replication may be independent of the attenuation of symptoms. When transformed tobacco plants were inoculated with tomato aspermy (cucumo)virus (TAV), satellite-producing plants failed to show the typical disease symptoms of TAV but produced as much viral RNA in the systemically infected leaves as did control plants which expressed symptoms but did not produce satellite RNA (Harrison et al 1987). The satellite-transformed plants infected with TAV also produced just as much satellite RNA as if they were infected with CMV. It therefore seems likely that the sattelite RNA is able to interfere directly with the ability of the virus to induce symptom formation as well as, at least with CMV infections, causing reduced accumulation of virus.

These results illustrate the potential usefulness of methods based on transformation for the production of plants which transcribe satellite RNA from the nuclear genome and which can thereby counteract the ability of virus to produce symptoms. Some further development is required before this approach can be applied in agriculture and careful comparison needs to be made with alternative strategies for achieving resistance/tolerance to viral disease. These strategies might include the expression of coat protein in transformed plants and the possible use of antisense RNA.

Antisense RNA of tobacco rattle virus

It was established several years ago with bacterial systems that RNA-mediated functions could be inhibited if the cells were engineered to contain the RNA complement of the biologically active form (Pines & Inouye 1986). Subsequently the same principle was tested successfully in eukaryotic cells, and there have been various demonstrations of inhibited gene expression caused by the production of antisense RNA (e.g. Kim & Wold 1985). Precisely how this inhibition comes about is not clear, and it may be that there are different types of inhibition in various systems. For example, in bacterial cells the most effective antisense RNA molecules were those complementary to the translational start region, implying an effect on the initiation of translation (Pines & Inouye 1986). However, it was found that 3′ terminal sequences were inhibitory (Kim & Wold 1985) in a eukaryotic system where the anti-sense effect was derived from a nuclear process. It is a fairly obvious extra-polation from these results to ask whether antisense inhibition could be used in plant cells to control viral infections. An attraction of this approach would be that, if successful, the same strategy would be applicable to many RNA viruses.

In our experiments we are using tobacco rattle virus (TRV) as a model in which to test the antisense strategy. This is an RNA virus with a bipartite genome in which the basic viral functions essential for replication and pathogenesis are encoded on RNA-1 (Harrison & Robinson 1986). Since it is clear from several systems that antisense inhibition depends on an excess of antisense over sense RNA, it is also clear that antiviral control will be achieved only if replicase function or production is blocked. For this reason our experiments have focused on RNA-1.

Several constructions were made in which the four viral genes of TRV RNA-1 (Boccara et al 1986) were each targeted by antisense RNA (Fig. 2). The various constructions were each inserted into the Ti plasmid-based ex-pression plasmid described above, in which the promoter is the 35S RNA promoter of CaMV. The constitutive activity of this promoter was considered especially important, because the use of developmentally regulated promo-ters would allow virus to replicate in cells where the promoter was not active. Presumably the virus could then enter adjacent cells where the antisense RNA was produced. However, it is likely that the amount of viral RNA originating from these sites of initial infection may be enough to overcome any antisense effect.

Despite these considerations, when transformed plants expressing the various antisense RNAs were inoculated with TRV there were no fewer lesions than in control plants. Nor was any effect observed on the ability of virus to spread within the transformed plant. We suggest that several reasons might account for this result:

(i) The type of sequence used as antisense RNA
The constructions analysed so far do not include the extreme 5' and 3' termini of TRV RNA-1. It is conceivable that these regions may be subject to antisense inhibition through their role in translation and viral RNA replication.

(ii) The intrinsic leakiness of antisense inhibition
Kim & Wold (1985) reported that antisense inhibition of thymidine kinase activity in animals leaves a residual activity of 10–20% of the control level. This degree of inhibition would be inadequate in a viral system, because this incomplete effect would allow the replication of sense RNA until an excess was obtained.

(iii) The effect of encapsidation
Viral RNA enters into and is maintained in the cell as encapsidated RNA

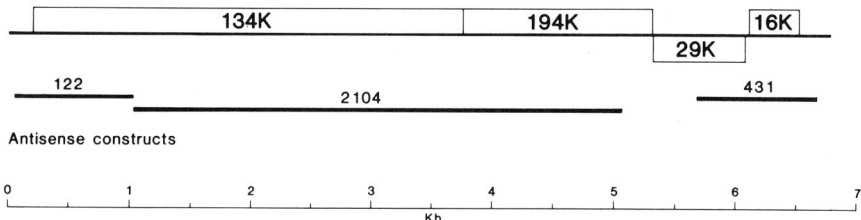

FIG. 2.   Antisense RNAs expressed in transformed plants. The four proteins coded for by TRV RNA-1 are indicated by large boxes. It should be noted that the 194K protein arises by readthrough of the 134K protein (W.D.O. Hamilton et al, unpublished). The black bars below define the regions of viral cDNA that were subcloned into *Rok*1 and subsequently transformed into *Nicotiana tabacum* using *Agrobacterium tumefaciens* and the Ti plasmid vector *Rok*1 (Fig. 1).

molecules. For structural reasons it is unlikely that encapsidated RNA will be accessible to antisense inhibition. However, as Wilson & Shaw (1985) have shown with several rod-shaped and icosahedral viruses, the virus particles themselves can serve as a template for translation. The coat protein is removed from the virion as the RNA is translated and structures consisting of ribosomes and partially disassembled virions have been visualized by electron microscopy.

Our experiments with TRV will allow the effect of encapsidation to be evaluated. Isolates of TRV which lack RNA-2 which encodes the capsid protein, but nevertheless retain the ability to infect systemically, will be inoculated onto tobacco plants expressing the various antisense RNA species. However, even if these experiments do show a positive effect of antisense RNA it is unlikely that antisense inhibition will be a useful antiviral strategy in agriculture because the majority of important viral infections involve encapsidated viral RNA.

In the light of these results, we consider that the best strategies for genetic engineering of resistance or tolerance to viral infection will be based on the satellite RNA approach and the use of coat protein genes, as described by Powell Abel et al (1986) and by Beachy et al (this volume).

A comparison of resistance/tolerance produced by coat protein expression and satellite RNA expression

The degree of protection

Tobacco plants expressing tobacco mosaic virus (TMV) coat protein showed most complete protection early after viral inoculation and if the viral inoculum was dilute. Plants lacking symptoms were free of virus (Powell Abel et al

1986). By contrast, the satellite-protected plants showed attenuation of disease with both dilute and concentrated viral inocula. All tobacco plants producing satellite were protected from the symptoms of CMV disease at least until flowering at 14 weeks after inoculation (Harrison et al 1987). However, the symptom-free leaves contained substantial amounts of virus – up to 20% of the amount in non-protected leaves after CMV infection, and an unreduced amount after TAV infection.

These features may have implications for the agricultural application of these approaches. For example, the dosage dependence of coat protein-mediated protection may limit the use of this approach to situations where infection pressure is low. With plants protected by expression of satellite RNA, the presence of high concentrations of virus in systemically infected leaves could serve as a reservoir of inoculum for nearby plants which are not resistant. It may be necessary, therefore, to limit use of satellite-protected plants to situations where spread of virus to nearby plants will not occur or where the adjacent crops are not susceptible to damage by the viral disease.

Relevance to viruses other than TMV and CMV

The use of coat protein expression is effective for TMV and alfalfa mosaic virus and, since all viruses have coat proteins, it seems possible that the same strategy will prove generally applicable. There is known to be a coat protein-mediated cross-protection with CMV (Dodds et al 1985), as with TMV, and CMV is a good candidate in which to test the general applicability of protection through coat protein expression.

Satellite RNA is a less widespread feature of plant viruses than coat protein. However, it has been possible to produce virus resistance in transgenic plants by expression of symptom-attenuating satellite RNA from tobacco ringspot virus (Gerlach et al 1987) as well as from CMV. In the future, when the principles of satellite-based protection are better understood, it may also prove possible to extend the range using artificial satellite sequences constructed *in vitro*.

Possible adverse effects

It is always possible to conjure up complex scenarios in which the use of genetically engineered plants in the field has severe adverse effects. Realistically, it is difficult to envisage adverse consequences of the expression of coat protein genes in plants. However, with satellite RNA the situation is more complicated. There are some virulent isolates which cause distinctive and severe disease when inoculated together with CMV. The symptoms may include white or yellow leaf mosaics and systemic necrosis (Kaper & Waterworth 1977), depending on the host plant and the particular satellite isolate.

The parts of the satellite sequence which determine the virulence property have not been identified, but they are likely to be quite short, because the nucleotide sequences of benign and virulent forms of the satellite differ in only a few residues (Gordon & Symons 1983). In order to avoid problems which may arise from the potential of CMV satellite RNA to cause disease we are now attempting to identify which parts of the satellite determine symptom production. Identification of the sequences which affect packaging into virions may be useful as well, since the satellite RNA can only be transmitted by aphids in an encapsidated form. Modification of either or both of these regions may then allow a 'safe' satellite to be used in the transformation work.

Concluding remarks

The experiments with satellite RNA described here and those with viral coat protein (Powell Abel et al 1986) illustrate the potential of transformation strategies for engineering resistance to viral infection. In particular they show that progress does not depend on isolating plant resistance genes — a difficult process in the absence of biochemical or genetical tags on the gene or its product. The successful examples, involving satellite RNA and viral coat protein, have both used the principle of turning the virus against itself, via the expression of viral sequence in the transformed plant. In future it may be possible to extend the range of viral sequences that can be used in this way by applying information derived from new methods of analysis of plant viral genomes. Examples of these methods include directed recombination and mutation analysis of both DNA (Daubert et al 1985) and RNA (Ishikawa et al 1986) viral genomes. Perhaps these future experiments, together with those described in this paper, will lead to the production of virus-resistant plants for use in agriculture.

Acknowledgement

We are grateful to the Rockefeller Foundation for support of our published and unpublished work described here.

References

Baulcombe DC, Saunders GR, Bevan MW, Mayo MA, Harrison BD 1986 Expression of biologically active viral satellite RNA from the nuclear genome of transformed plants. Nature (Lond) 321:446–449

Beachy RN, Powell Abel P, Nelson RS, Register J, Tumer N, Fraley RT 1987 Genetic engineering of plants for protection against virus diseases. In: Plant resistance to viruses. Wiley, Chichester (Ciba Found Symp 133) p 151–169

Boccara M, Hamilton WDO, Baulcombe DC 1986 The organisation and interviral homologies of genes at the 3' end of tobacco rattle virus RNA 1. EMBO (Eur Mol Biol Organ) J 5:223–229

Daubert SD, Schoelz J, Debao Li, Shepherd RJ 1984 Expression of disease symptoms in cauliflower mosaic virus genomic hybrids. J Mol Appl Genetics 2:537–547

Dodds JA, Lee SQ, Tiffany M 1985 Cross protection between strains of cucumber mosaic virus: effect of host and type of inoculum on accumulation of virions and double-stranded RNA of the challenge strain. Virology 144:301–309

Gerlach WL, Llewellyn D, Haseloff J 1987 Construction of a plant disease resistance gene from the satellite RNA of tobacco ringspot virus. Nature (Lond) 328:802–805

Gordon KHJ, Symons RH 1983 Satellite RNA of cucumber mosaic virus forms a secondary structure with partial 3'-terminal homology to genomal RNAs. Nucl Acids Res 11:947–960

Harrison BD, Robinson DJ 1986 Tobraviruses. In: Van Regenmortel MHV, Fraenkel-Conrat H (eds) The plant viruses. Plenum, New York, vol 2:339–369

Harrison BD, Mayo MA, Baulcombe DC 1987 Virus resistance in transgenic plants that express cucumber mosaic virus satellite RNA. Nature (Lond) 328:799–802

Ishikawa M, Meshi T, Motoyoshi F, Takamatsu N, Okada Y 1986 In vitro mutagenesis of the putative replicase genes of tobacco mosaic virus. Nucl Acids Res 14:8291–8305

Jacquemond M, Lot H 1981 L'ARN satellite du virus de la mosaïque du concombre. I. Comparison de l'aptitude à induire la nécrose de la tomate d'ARN satellites isolés de plusieurs souches du virus. Agronomie (Paris) 1:927–932

Kaper JM, Waterworth HE 1977 Cucumber mosaic virus associated RNA5: causal agent for tomato necrosis. Science (Wash DC) 196:429–431

Kaper JM, Waterworth HE 1981 Cucumoviruses. In: Kurstak E (ed) Handbook of plant virus infections and comparative diagnosis. Elsevier/North-Holland Biomedical Press, Amsterdam, p 257–332

Kim SK, Wold BJ 1985 Stable reduction of thymidine kinase activity in cells expressing high levels of anti-sense RNA. Cell 42:129–138

Lot H, Jonard G, Richards K 1977 Cucumber mosaic virus RNA 5. Partial characterization and evidence for no large sequence homologies with genomic RNAs. FEBS (Fed Eur Biochem Soc) Lett 80:395–400

Mossop DW, Francki RIB 1979 Comparative studies on two satellite RNAs of cucumber mosaic virus. Virology 95:395–404

Piazzolla P, Tousignant ME, Kaper JM 1982 Cucumber mosaic virus-associated RNA 5. IX. The overtaking of viral RNA synthesis by CARNA 5 and dsCARNA 5 in tobacco. Virology 122:147–157

Pines O, Inouye M 1986 Anti-sense RNA regulation in prokaryotes. Trends Genet 2:284–287

Powell Abel P, Nelson RS, De B et al 1986 Delay of disease development in transgenic plants that express the tobacco mosaic virus coat protein gene. Science (Wash DC) 232:738–743

Tien Po, Zhanz Xiuhua, Qiu Binsheng et al 1987 Satellite RNA as a biological control agent of plant diseases caused by cucumber mosaic virus. Ann Appl Biol 111:143–152

Wilson TMA, Shaw JG 1985 Does TMV uncoat cotranslationally in vivo? Trends Biochem Sci 10:57–60

DISCUSSION

Dodds: Dr Baulcombe, what happens to your transgenic plants if, rather than inoculate them with selected strains of CMV that do not have satellite, you try to infect them with a virulent isolate of CMV where the basis for virulence is the presence of one of the more aggressive CARNA-5 satellites?

Baulcombe: We hoped that we would be able to get cross-protection by the endogenous satellite against the virulent satellites. We haven't been able to achieve that yet. We infected pT104 transformed and pT105 transformed tobacco plants with a CMV isolate containing the Y satellite. The Y satellite on tobacco produces a spectacular yellow chlorosis and mosaic. It can be distinguished from the satellite derived from the transcripts on the basis of size: the Y satellite is slightly larger. Typical Y satellite symptoms were observed in both the pT104 and pT105 plants, which was disappointing. After 21 days there is more of the Y satellite than of the endogenous satellite in the pT104 plants. However, the pT105 plants have more endogenous satellite than Y satellite. The obvious extrapolation is to use higher-order multimers than the 2.3-mer and to see whether we can extend this to such a level that we obtain almost complete competition against the Y satellite.

Goldbach: For this satellite approach one has to transform more than one copy which results in repeated sequences in the plant chromosome. Do you expect, or did you find, any instability of the inserted DNA? One might expect recombinations between the multimeric sequences of the satellites.

Baulcombe: We haven't observed any recombinations. We have looked at a dozen different lines at the RNA level that are making satellites and they all seem to make transcripts of correct size. We have not observed significant instability.

Beachy: This might be a good system in which to look for RNA–RNA recombination—for example, between transcribed satellite and any input satellite from an infection. What do you think about the possibility of recombinational instability of the satellite in the face of infection with another strain?

Baulcombe: It is possible. We do not know much about the frequency of RNA–RNA recombination events in the absence of any selection for them. I would imagine they are quite rare.

Beachy: From what we have learned from this system, and from animal viruses, would it be possible to engineer any plant virus to be a DI particle? Is it feasible to try to make a DI particle out of TMV?

Baulcombe: I certainly think it is feasible. The starting elements are very easy to envisage. One would simply make a construct using the 5' and the 3' termini of a viral sequence. Perhaps the transcripts would be replicated without interfering with symptom production. If it doesn't affect symptom production, one will have to insert something that will ensure that it would interfere either with symptom production or with the ability of the virus to replicate. There are several types of sequence that, when introduced in that way, might interfere with the virus.

van Vloten-Doting: Making interfering particles may be more difficult. There is an internal sequence in brome mosaic virus and alfalfa mosaic virus, at least in RNA-3, that also seems to be essential. This may even apply to RNA 1 and 2 because there are some indications that there are internal binding sites for coat protein which, at least for AlMV, are essential. This probably applies to other viruses. It is surprising that we have not seen interfering particles in natural plants, whereas the phenomenon is easily detected with animal viruses. When we inoculate plants we normally do so with mass infections which would be good for selecting for defective interfering particles.

Baulcombe: The need to introduce internal sequences would be a complicating feature but it needn't necessarily deter us from this approach.

Bruening: We have some results which are relevant to your observations that replication of the satellite and the satellite's ability to protect the plant against symptoms and reduce virus replication may not be connected. We have used the satellite of tobacco ringspot virus to protect cowpea plants against, not tobacco ringspot virus, but cherry leafroll virus, another nepovirus. The yield of cherry leafroll virus was reduced if the mass-based inoculum levels of this virus and the satellite RNA were similar. The plants were protected against the worst symptoms. This occurs even though cherry leafroll virus does not support the replication and encapsidation of the tobacco ringspot virus satellite RNA. In other experiments we used more concentrated inocula of cherry leafroll virus. There was no diminution of viral replication, but the cowpeas were nevertheless protected against the most severe symptoms that this virus would otherwise induce.

van Vloten-Doting: Does that mean that virus replication and symptom expression are often unrelated?

Baulcombe: There must be a minimum level of virus required for symptom production but above that minimum level, the amount of virus may not be important.

Bruening: Other factors, such as interaction with minor gene products of the virus, will prevent symptom production even when virus increase is considerable.

Baulcombe: The tomato aspermy virus data that I described also demonstrate that point. Tobacco plants that are infected with TAV and that produce satellite RNA are more or less symptom free, despite the fact that they are full of virus.

Harrison: We investigated whether the satellite-transformed plants are protected against the effects of a range of different viruses including one other cucumovirus, peanut stunt. This does not work and peanut stunt infection does not induce replication of the satellite. Nor does infection with viruses in several other groups. Thus, there does seem to be a strong virus specificity.

Dr Baulcombe mentioned that the attenuation of symptoms lasts for a long time after CMV infection. In fact we have kept infected satellite-transformed plants until flowering and they still show virtually no symptoms.

We have also inoculated the virus by vector aphids, which is the natural mode of spread. pT104-transformed and control seedlings were equally readily infected using two aphids per plants. However, the satellite-transformed seedlings infected by aphids were well protected from the effects of infection and grew almost as strongly as virus-free plants. To test infected satellite-transformed plants as sources of virus, we fed aphids on leaves of pT104-transformed and of control plants that were infected with CMV and then transferred the aphids to *Nicotiana benthamiana* seedlings. In general there was less transmission by the aphids when the virus came from the satellite-transformed plants, except shortly after infection. There is a delay before the satellite exerts its effect and during this period there was substantial multiplication of CMV. It is only after a few systemically infected leaves have been produced that viral replication is suppressed. In general the satellite-transformed plants were much poorer sources of virus than the controls.

Dodds: Is that isolate attenuated because it has satellite associated with it? Is it not an intrinsic change in CMV?

Harrison: No, it is because it has acquired satellite RNA.

van Vloten-Doting: Do the results also hold for tomato aspermy virus? In the source plant there would be a lot of virus: one would expect efficient transmission and infection to occur regardless of whether the satellite RNA was present.

Harrison: Yes; satellite-transformed plants infected with tomato aspermy virus are only slightly poorer virus sources for aphids than control plants.

Dodds: The activation and encapsidation of CARNA-5 in transgenic plants after natural infection with CMV seems to be a guaranteed mechanism for disseminating satellite RNA by aphid transmission to places where it might not have initially been. Is that a good thing or a bad thing?

Baulcombe: We would concede that the production of potentially infectious RNA in transformed plants is generally a bad thing. We are attempting to disable the satellite so that it can be replicated in the plant by the virus but not transmitted subsequently. We don't know whether or not that will be possible.

Harrison: Dr van Emmelo, would you like to add to what Dr Baulcombe has said about the antisense approach to virus resistance?

van Emmelo: An important project at Plant Genetic Systems involves using genetic manipulation techniques to obtain sugarbeet plants resistant to beet necrotic yellow vein virus. This virus causes rhizomania disease, which results in a remarkable reduction not only in the size of the sugarbeet but also in its sugar content.

A feasibility study was started two years ago, using tobacco–tobacco mosaic virus as a model system. Antisense technology had recently produced some encouraging results, in both prokaryotic and eukaryotic systems. Our goal was to inhibit one of the first steps in virus multiplication, the production of the replicase protein, by the constitutive presence of antisense-containing mRNAs in the plant. The target chosen for interaction was a region of some 30 nucleotides surrounding the replicase ribosome binding site.

Chemically synthesized oligodeoxynucleotides were introduced in TMV RNA-directed wheat germ translation systems. When the ratios between oligonucleotide and target were as high as 50, synthesis of the corresponding viral protein almost ceased. Other oligodeoxynucleotides, such as those complementary to the extreme 5' end, were tested for their inhibitory effect in the same system, but were less efficient in interfering with protein synthesis. This does not seem to be a general feature, because different results were obtained when other viral RNAs were used to direct synthesis. This indicates that before starting to do time-consuming work in plants it is advisable to do *in vitro* experiments in order to select for the best interfering antisense molecule.

For the *in vivo* constructions we assembled cassettes consisting of the *NPTII* gene, giving resistance to kanamycin, followed at its 3' end by one copy, or three copies, of the antisense fragment with the 30 nucleotides complementary to the ribosome binding site, or a copy without that 30 nucleotide region, and an appropriate polyadenylation signal. When this cassette was placed under the control of the bacteriophage SP6 promoter, transcripts of some 1200 nucleotides were obtained *in vitro*. These transcripts interacted with the TMV RNA in such a way that the complexes between target and interfering agent, although only complementary over a region of 30 nucleotides, could be detected by electrophoresis on agarose gels. The effect of these long transcripts in a wheat germ translation system was impossible to assay reproducibly, because it is nearly impossible to achieve the high concentrations of large polynucleotides needed for positive interference in a wheat germ system.

The hybrid DNA molecules were coupled to different plant promoters, the strongest being the ribulose bisphosphate carboxylase (RUBISCO) small subunit promoter, isolated from *Arabidopsis*. By means of *Agrobacterium tumefaciens* leaf disc transformation we were able to obtain many tobacco (SR1 or Samsun) shoots which were highly resistant to the antibiotic kanamycin. Plants grown from these shoots were assayed for resistance to TMV. Whereas the control plants, without antisense in the *NPTII* messenger, were all fully infected by the virus, many transgenic plants that produced a number of neomycin phosphotransferase messengers showed a remarkable delay, not only in the appearance of symptoms, but also in viral multiplication (judged by ELISA and Western blotting followed by immunodetection).

When *N. tabacum* Xanthi nc plants were transformed in the same way, no differences between antisense-producing and control plants were detected in terms of the number or size of the lesions. When these plants were transferred to 32°C (four days after the mechanical inoculation), systemic spread of the virus was almost completely inhibited in the first population. This was assayed by taking young leaves two weeks after transfer to the higher temperature and using ELISA to detect the virus.

Beachy: Did you use seedling progeny of the transformed plants?

van Emmelo: All the experiments were done using plants derived from the original transformed shoots. We have now obtained seeds and we shall examine the seedlings soon.

Beachy: Care must be taken when using clonally propagated cuttings. We also experienced some variability in our early studies using rooted cuttings. We always had resistance but we were not sure what it meant. As we went to the next generation with small three-leaf seedlings, the level of inoculum that the plant tolerated was considerably less.

van Emmelo: It is certainly surprising that some of the plants expressing a lot of messengers show symptoms, whereas others with a low amount of antisense do not seem to.

Harrison: What was the inoculum concentration in these tests?

van Emmelo: 1 µg/ml, with abrasive.

Harrison: So it is a heavy inoculum.

van Vloten-Doting: Do you inoculate and then after a certain time you remove the inoculated leaves?

van Emmelo: We didn't remove the inoculated leaves. In the Samsun experiment we studied leaves just above the inoculated one, whereas in the Xanthi experiment we looked at the top leaves, 14 days after transfer to 32 °C.

Hohn: You were looking for resistance related to a DNA sequence capable of producing antisense RNA. Where you did not see any effect, did you investigate whether antisense RNAs were really present?

van Emmelo: We demonstrated that the antisense RNA was present even where infection occurs.

Beachy: Your construct showed the 30-mer at the 3′ end. Did you try the 30-mer as the 5′ non-translated region? Did you think there might be a greater chance that it would be non-folded if the 30-mer was the 5′ end of the messenger rather than the 3′ end? One might expect the 5′ end to be more accessible to annealing than the 3′ end.

van Emmelo: There is translation from the *NPTII* messenger, so the 5′ part of the messenger might be expected to be unavailable for interaction, if we accept the Kozak rules (Kozak 1981). That is why we put it at the 3′ end. Even that does not completely guarantee that ribosomes won't interfere in some way with the antisense binding.

Zimmern: Have you tried these experiments either *in vitro* or in field trials with other strains of TMV? If you were depending on hybridization with a short sequence, you might not get protection against many strains.

van Emmelo: Those are experiments we want to do as soon as possible. We shall use different TMV strains, or an antisense where different mutations are artificially introduced. It will be important to detect the ratio needed to interfere with translation in the different cases.

Zimmern: Then you would have to put in a series of different antisense RNAs to get protection against a spectrum of viruses.

van Emmelo: That would be one of the advantages of the antisense technology— obtaining resistance against a lot of unrelated viruses with a single construction.

Matthews: Is it possible to improve the degree of control by incorporating an antisense transformer at both the 5′ and the 3′ ends?

van Emmelo: Dr Mike Koziel did similar experiments to mine. He combined 90 nucleotides from the 5′ end with 40 nucleotides from the 3′ end, making a very strange construction. He also obtained the same type of response. I have the impression that his results were better than ours. Few of his plants showed symptoms and he could not detect TMV. However, there were some exceptions.

Reference

Kozak M 1981 Possible role of flanking nucleotides in recognition of the AUG initiator codon by eukaryotic ribosomes. Nucl Acids Res 9:5233–5252

Plant DNA viruses as gene vectors

B. Hohn, N. Grimsley, B. Pisan and T. Hohn

Friedrich-Miescher-Institut CH-4002, Basel, Switzerland

Abstract. Caulimoviruses and geminiviruses are the only known plant DNA viruses. Both groups are candidates for carrying foreign DNA into plants, spreading it systemically and expressing high yields of the corresponding gene product. This has been achieved with hybrids of cauliflower mosaic virus (CaMV) and certain model genes. A major obstacle to the use of this technology is the high mutation rate of CaMV, probably caused by its mode of replication via reverse transcription, involving switches of nascent DNA strands on single-stranded RNA and DNA templates. If these occur at illegal positions, deletions and duplications are created. These are rarely observed with wild-type infections but deletions of foreign sequences are selected for if the total length of the hybrid genome is too large, when inserted sequences interfere with virus transcription and translation, when the secondary structure of replicative intermediates is changed, or if expressed payload protein is disadvantageous to virus or plant cell. Similar problems arise with geminiviruses and single-stranded RNA viruses with single-stranded genomic replicative intermediates in their life cycles. This problem of instability could be avoided by creating master copies of double-stranded DNA of the hybrid virus in the plant cells from which the single-stranded replicative intermediates are produced continuously. This could be achieved by agroinfection (transfer of virus genomes as double-stranded DNA multimers into the host cell with the help of agrobacteria). An interesting achievement in this field is agroinfection with the geminivirus, maize streak virus.

1987 Plant resistance to viruses. Wiley, Chichester (Ciba Foundation Symposium 133) p 185–195

There are only two known types of plant DNA viruses: the caulimoviruses, containing a double-stranded (ds) DNA molecule of 8000 base pairs; and the geminiviruses, which exist as monopartite and bipartite classes with about 2700 nucleotides (nt) of single-stranded circular DNA packaged per particle (Howell 1986). Both types of viruses are usually transmitted by insects and the infection spreads systemically from the originally affected tissue to the newly developing leaves. Both are candidates for gene vectors in plants. Caulimoviruses have been studied most so far. However, in the future there may be more interest in geminiviruses because some of them infect Gramineae. Neither caulimovirus nor geminivirus DNA integrates into the host genome during their natural life cycles; thus their main uses would be as expression vectors and as vehicles for fast systemic amplification of a given coding sequence. Heritable modifications of plants are more appropriately effected by direct and *Agrobacterium*-mediated transformation (Potrykus et al 1987, Rogers & Klee 1987). In this paper we shall concentrate mainly on cauliflower mosaic virus (CaMV), the type member of caulimo-viruses.

Properties of cauliflower mosaic virus in relation to its use as a plant vector

CaMV exists in different forms. Supercoiled 8000 bp circular DNA is present in the nuclei of infected plants. Transcription of this yields an RNA species with a 180 nt terminal repeat, because the host RNA polymerase II selectively bypasses its termination signal on the first round of RNA synthesis. This '35S RNA' resembles in structure and creation the viral RNA of retroviruses, which also have terminal repeats (R-regions). CaMV particles, finally, contain open circular double-stranded DNA derived from the genomic RNA by reverse transcription using Met-tRNA as a primer for $(-)$DNA and oligo(G) stretches for $(+)$DNA. Reverse transcription occurs in typical viral inclusion bodies, large proteinaceous structures occupying a major portion of the infected cytoplasm. Packaged DNA has short single-strand overlaps at the position of the primer binding sites and priming sites that are obviously remnants of the reverse transcription process and only repairable and ligatable upon re-infection of a nucleus (reviews: T. Hohn et al 1985, Mason et al 1987).

CaMV translation occurs in the inclusion bodies also: thin sections of certain classes of these organelles reveal ribosomes attached to their surface (Shepherd et al 1979). Total viral RNA isolated from infected cells shows only limited activity in *in vitro* translation systems. This might be due to the need for the inclusion body environment and/or other virus-and host-encoded factors not present in the wheat germ and reticulocyte systems. The only translation product obtained in reasonable quantities *in vitro* is inclusion body matrix protein, which originates from a second CaMV RNA species, the 19S RNA (Fig. 1). No hard evidence for additional RNA species exists and the genomic 35S RNA is assumed to be the mRNA for the other five to seven CaMV primary gene products, some of which will be processed to increase the variety of CaMV-derived proteins. These genes are tightly arranged with only -1, 1 or a few bases intergenic distance. Nonsense and frameshift mutations in the non-essential open reading frames (ORFs) VII and II are polar if they either increase the intergenic distance or cause an ORF to extend beyond the ATG of the ORF normally following. This has led to the proposal of a polycistronic type of translation, whereby the ribosome does not fall off after each stop codon but resumes 'relay race translation' at the nearest start codon (Sieg & Gronenborn 1982, Dixon & Hohn 1984). Additional mechanisms, such as selective ATG bypass, might have to be proposed for translation of the most abundant CaMV capsid gene which is coded for by the ORF IV on the 35S RNA. The poor translatability of ORF VII, the first ORF on 35S

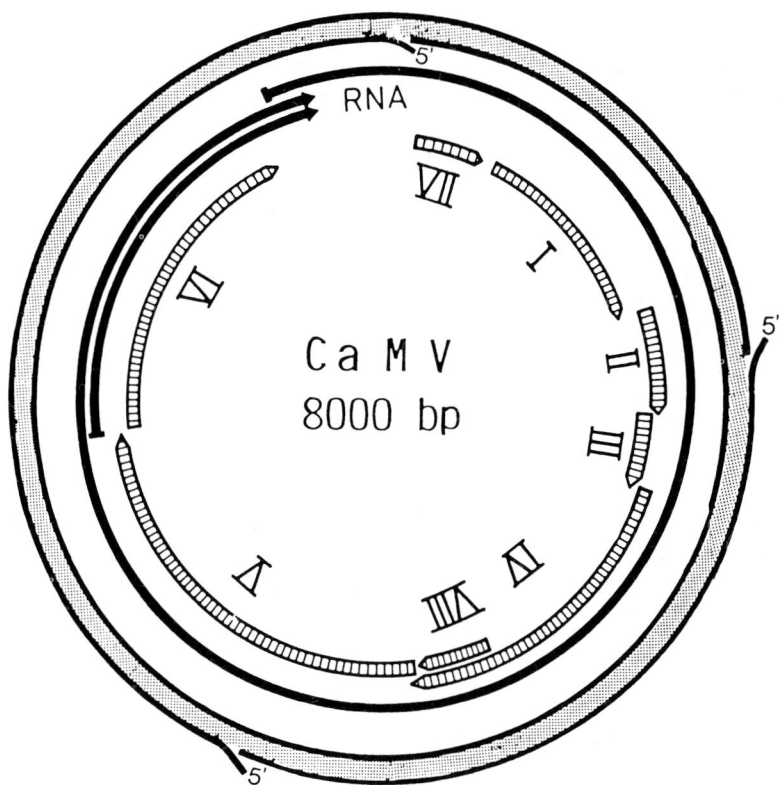

FIG. 1.   Map of cauliflower mosaic virus showing the double-stranded DNA with its typical single-strand overhangs, the two coterminal transcripts, and the open reading frames (ORFs) I–VIII. Possible functions for the ORFs include: II Aphid transmission factor. IV (*gag*). N terminal: coat protein; C terminal: basic protein. V (*pol*). N terminal: protease; C terminal: reverse transcriptase. VI Inclusion body matrix protein and host range factor (*env*). VII and VIII Basic proteins (regulators?).

RNA *in vitro*, is highly improved by removal of the 600 nt long leader. Likewise, the translatability of ORF I, the second ORF, is improved by the removal of the leader and of ORF VII, and so forth (Gordon et al 1987). Preliminary results using transient expression in non-host (*Nicotiana tabacum*) and host (*Orychophragmus violacea*) protoplasts indicate that, *in vivo*, the leader is bypassed, possibly with the help of a host-encoded factor, and that virus-encoded factors may be necessary for relay race translation (Bonneville et al 1987).

CaMV shows a very high recombination rate (Lebeurier et al 1982). This might be due to the mode of replication. The reverse transcriptase has to switch templates when reaching the 5' terminus of the genomic RNA during

(−) strand DNA synthesis and again when reaching the 5' terminus of (−)DNA during (+) strand DNA synthesis. It is conceivable that *inter*-genomic template switches are the cause of recombination between pairs of mutants, interfering with any complementation experiments (Lebeurier et al 1982, Howell et al 1981). Other 'illegal' template switches between homologous molecules might be the reason for deletions resulting in the accumulation of small CaMV satellite circles in nuclei after infection (Olszewski & Guilfoyle 1983), for the occurrence of deletions in the insect transmissibility gene after several cycles of mechanical inoculation (Howarth et al 1981), and even for duplications in CaMV DNA subpopulations (Fang et al 1985, Penswick et al 1987a).

Developing CaMV into a plant gene vector

A precondition for developing a viral genome into a vector is a method for re-introducing manipulated DNA into the host. CaMV and its DNA can enter the host plants after mechanical inoculation of the leaf by rubbing it with the inoculum and silica powder (Hull & Shepherd 1977). The absence of the short single-strand overlaps in cloned DNA does not impair infectivity (Lebeurier et al 1980, Howell et al 1980). However, cloned DNA either has to be excised from the bacterial plasmid — the linear molecule will be circularized in the host plant — or must be present as a multimer or at least as an entity with terminal repeats. From the latter the plant will recombine unit-length CaMV DNA molecules or produce directly genomic RNA if the promoter/terminator region is repeated. The infectivity of cloned DNA allowed testing of mutants produced *in vitro* (Dixon et al 1983, Daubert et al 1983), which showed that all the CaMV ORFs, except II and VII, are indispensable for systemic virus spreading. Polar nonsense and frameshift mutations of ORFs VII and II are unstable and revert by second site mutations (Sieg & Gronenborn 1982, Dixon & Hohn 1984). However, complete deletions of these ORFs are viable and stable, making space for an 800 bp payload. A single *Xho*I restriction site had been used for small viable insertions into ORF II (Gronenborn et al 1981); subsequently developed CaMV vectors used this restriction site but deleted all other ORF II sequences. Such a vector finally allowed the successful cloning of methotrexate-resistant bacterial dihydrofolate reductase in plants, yielding high levels of the enzyme (review: Brisson & Hohn 1986).

Limitations of CaMV vectors

Experiments designed to investigate further uses of the CaMV vector quickly showed that inserted genes of interest have to be adapted to the peculiarities of CaMV proliferation. The high recombination (deletion) rate in CaMV will create variant genomes that compete with the original construct for proliferation. Inserts might be removed, shortened or altered in the progeny virus population if they are too large to be packaged into the CaMV capsid; if they are too G-rich (for example, bacterial neomycin phosphotransferase) to become smoothly reverse-transcribed; if they contain transcriptional stop signals interfering with genomic RNA production; if they code for products which might interfere with the virus life cycle or host metabolism; and if the rules of proper arrangement of ORFs are disobeyed, leading to interference with the peculiar CaMV translation mechanism. In addition, introns will be removed during transport of the genomic RNA from nucleus to cytoplasm (B. Hohn et al 1986). Thus systemic CaMV expression vectors and the inserted coding regions have to be tailored exactly to create the most natural transitions between the ORFs (Brisson et al 1984). Exact replacement of ORF II has been more successful; replacements of ORF VII show more instability, possibly because of the initiation of translation at that position (Penswick et al 1987b). Although these constraints originally made work with CaMV vectors tedious, improvements of DNA sequencing and *in vitro* mutagenesis technology using synthetic oligonucleotides now allow the construction of hybrid CaMV vectors with relative ease (Penswick et al 1987b).

Applications of CaMV vectors

Several practical uses of the CaMV vector can be envisaged. Mature plants could be made resistant to toxic agents, as the model experiment with methotrexate resistance (Brisson et al 1984) and experiments with the Chinese hamster metallothionein gene (Lefebvre & Laliberté 1987) have shown. Edible plants could be used for the production of substances of interest and, in contrast to those cloned in microbes, these could be used without tedious purification because toxins are absent. In this way the human α-interferon (Penswick et al 1987b) and the flounder anti-freeze genes (A. Peterhans & J. Paszkowski, unpublished results) were successfully cloned and maintained and the corresponding proteins expressed in turnips for several weeks. There are further applications for basic research. For instance, tissue-specific activators or repressors could be tested in the whole plant after infection with hybrid CaMV containing the corresponding control gene. Special studies involving sequences cloned in CaMV vectors have helped to answer questions about priming sites for reverse transcription (Pietrzak & Hohn 1985), splicing in plants (B. Hohn et al 1986), and processing of *Agrobacterium* T-DNA borders in plants (G. Bakkeren, Z. Koukolikovà-Nicola, N. Grimsley and B. Hohn, unpublished results).

Agroinfection

The genomic instability of autonomously replicating eukaryotic vectors is a problem not only for CaMV vectors but also for yeast plasmids and animal virus vectors. However, in animal systems high stability could be achieved with retroviral integrative vectors. In this case integration depends on the presence of specific integration sites on the vector and on an integrase encoded either by the viral genome or by a helper. CaMV does not have these sites and functions. However, a related situation to integrating retrovirus can be achieved with CaMV in combination with *Agrobacterium*-mediated plant transformation, or 'agroinfection' (review: Grimsley & Bisaro 1987). In this method virus DNA multimers, or at least monomers with appropriately designed terminal redundancy, are placed between the T-DNA borders of *Agrobacterium* Ti plasmids or their derivatives and transferred to plants using the *Agrobacterium* virulence system. Constructs are designed such that the terminal repeats of the inserted CaMV genome are retrovirus-like — they include the promoter/terminator region (Grimsley et al 1986). This allows the establishment of a stable master copy. Viral genomes can be constantly shed from this, can amplify via transcription/reverse transcription, and can spread systemically.

Agroinfection has additional interesting aspects:

(1) It provides a new strategy for making manipulated genomic DNA or cDNA of other plant viruses infectious where mechanical inoculation is not possible. Alternative strategies of *in vitro* packaging, as for bacteriophage λ, have not been developed for plant viruses. Experiments designed to load the insect vector with viral DNA have not, to our knowledge, led to plant transfection.

(2) Transgenic plants can be produced from protoplasts or leaf discs agro-infected with mutant viral strains produced *in vitro* that are lethal in normal viral infectivity tests and therefore cannot otherwise be easily studied. Such transgenic plants will accumulate replication and assembly intermediates and thus allow study of complex viral life cycles. Some of these transgenic plants may be cross-protected against superinfecting wild-type virus. Others might be useful for complementation of viral vectors which do not themselves encode all the information necessary for viral replication because they are deleted for large portions of their genome to provide increased payloads. Finally, transgenic 'non-host' plants can be used to study incompatible virus–host interactions.

Recently, agroinfection was applied to maize streak virus (MSV), a gemini-virus infecting grasses (Grimsley et al 1987). This virus is not infectious as free DNA, but only if transmitted as virus particles by leaf hoppers. Members of the Gramineae were not considered to be hosts for *Agrobacterium* because of the lack of tumour formation in these plants. Nevertheless, MSV could be transmitted to maize plants using *Agrobacterium* harbouring binary vectors with dimers of MSV DNA placed between the T-DNA borders. This technology opens up the possibility of developing and testing MSV as a vector for

Gramineae, studying its life cycle with the help of mutants, and possibly cross-protecting transgenic maize against the virus. MSV spreads systemically from the original point of infection, but insertion of the MSV genome into maize nuclear DNA has not yet been shown, and whether or not MSV constructs would have less tendency to rearrange in plants than has CaMV remains to be seen. Single-stranded DNA viruses, as well as RNA viruses, might not show better fidelity of replication than viruses using reverse transcriptase. However, better alternatives for plant virus vectors will not be available until true plant ds DNA viruses are found. On the other hand, the instability of the self-replicating constructs provides biological containment.

Acknowledgements

We wish to express our thanks to J. Davies for many valuable discussions, to C. Matsui for the *Orychophragmus violacea* cell line mentioned and to all our colleagues who provided unpublished information as cited.

References

Bonneville J-M, Fütterer J, Penswick J, Sanfaçon H, Pisan B, Hohn T 1987 Transient expression of CaMV DNA. In press

Brisson N, Hohn T 1986 Plant virus vectors: cauliflower mosaic virus. Methods Enzymol 118:659–668

Brisson N, Paszkowski J, Penswick JR, Gronenborn B, Potrykus I, Hohn T 1984 Expression of a bacterial gene in plants by using a viral vector. Nature (Lond) 310:511–514

Daubert S, Shepherd RJ, Gardner RC 1983 Insertional mutagenesis of the cauliflower mosaic virus genome. Gene (Amst) 25:201–208

Dixon L, Hohn T 1984 Initiation of translation of the cauliflower mosaic virus genome from a polycistronic mRNA: evidence from deletion mutagenesis. EMBO (Eur Mol Biol Organ) J 3:2731–2736

Dixon LK, Koenig I, Hohn T 1983 Mutagenesis of cauliflower mosaic virus. Gene (Amst) 25:189–199

Fang R, Wu X, Bu M, Tian Y, Cai F, Mang K 1985 Complete nucleotide sequence of CaMV (Xinjing isolate) genomic DNA. Chin J Virol 1:247–256

Gordon K, Pfeiffer P, Fütterer J, Hohn T 1987 Translation of cauliflower mosaic virus ORFs *in vitro*. In press

Grimsley N, Bisaro D 1987 Agroinfection. In: Hohn T, Schell J (eds) Plant DNA infectious agents. Springer, Wien & New York, p 87–107

Grimsley N, Hohn B, Hohn T, Walden R 1986 Agroinfection, a novel route for plant viral infection using Ti plasmid. Proc Natl Acad Sci USA 83:3282–3286

Grimsley N, Hohn T, Davies JW, Hohn B 1987 *Agrobacterium*-mediated delivery of infectious maize streak virus into maize plants. Nature (Lond) 325:177–179

Gronenborn B, Gardner RC, Schaefer S, Shepherd RJ 1981 Propagation of foreign DNA in plants using cauliflower mosaic virus as vector. Nature (Lond) 294:773–776

Hohn B, Balasz E, Rüegg D, Hohn T 1986 Splicing of an intervening sequence from hybrid CaMV RNA. EMBO (Eur Mol Biol Organ) J 5:2759–2762

Hohn T, Hohn B, Pfeiffer P 1985 Reverse transcription in a plant virus. Trends Biochem Sci 10:205–209

Howarth AJ, Gardner RC, Messing J, Shepherd RJ 1981 Nucleotide sequence of naturally occurring deletion mutants of cauliflower mosaic virus. Virology 112:678–685

Howell S 1986 The molecular biology of plant DNA viruses. CRC Crit Rev Plant Sci 2:287–316

Howell SH, Walker LL, Dudley RK 1980 Cloned cauliflower mosaic virus DNA infects turnips (*Brassica rapa*). Science (Wash DC) 208:1265–1267

Howell SH, Walker LL, Walden RM 1981 Rescue of in vitro generated mutants of cloned cauliflower mosaic virus genome in infected plants. Nature (Lond) 293:483–486

Hull R, Shepherd RJ 1977 The structure of cauliflower mosaic virus genome. Virology 79:216–230

Lebeurier G, Hirth L, Hohn T, Hohn B 1980 Infectivities of native and cloned cauliflower mosaic virus DNA. Gene (Amst) 12:139–146

Lebeurier G, Hirth L, Hohn B, Hohn T, 1982 In vivo recombination of cauliflower mosaic virus DNA. Proc Natl Acad Sci USA 79:2932–2936

Lefebvre DD, Laliberté J-F 1987 Expression of metallothionein in *Brassica campestris* using CaMV as a vector. In: Verma DPS, Brisson N (eds) Third Int Symp Mol Gen Plant-Microbe Interact. Martinus Nijhoff, New York, p 32–34

Mason WS, Taylor JM, Hull R 1987 Retroid virus genome replication. Adv Virus Res 32:35–96

Olszewski N, Guilfoyle T 1983 Nuclei purified from cauliflower mosaic virus-infected turnip leaves contain subgenomic, covalently closed circular cauliflower mosaic virus DNAs. Nucl Acids Res 11:8901–8914

Penswick J, Hübler R, Hohn TA 1987a Duplication in CaMV DNA separating the *gag* and *pol* coding regions.

Penswick J, Horrisberger M, Hohn T 1987b Expressing human interferon α in a plant using a CaMV vector.

Pietrzak M, Hohn T 1985 Replication of the cauliflower mosaic virus: studies on the putative plus strand primer binding site. Gene (Amst) 33:169–179

Potrykus I, Paskowski J, Shilito RD, Saul MW 1987 Direct gene transfer to plants. In: Hohn T, Schell J (eds) Plant DNA infectious agents. Springer, Wien & New York, p 229–247

Rogers SG, Klee H 1987 Pathways to plant genetic manipulation employing *Agrobacterium*. In: Hohn T, Schell J (eds) Plant DNA infectious agents. Springer, Wien & New York, 179–203

Shepherd RJ, Richins R, Shalla TA 1979 Isolation and properties of the inclusion bodies of cauliflower mosaic virus. Virology 102:389–400

Sieg K, Gronenborn B 1982 Evidence for polycistronic messenger RNA encoded by cauliflower mosaic virus. NATO Adv Study Inst Adv Course 1982:154

DISCUSSION

Goldbach: What is the maximum size of the DNA that can be inserted into the CaMV genome?

Hohn: At present, 1000 base pairs. With the help of this agroinfection it might be possible to produce artificial host plants that have some of the virus function and then one could load more information into the vector. However, I don't think that the size of an insert is the major problem. There are a lot of genes that one can use in that 1000 base size range. The major problems are the need to make very precise constructions and the fact that one can only use open reading frames. One cannot use gene constructs with their own promoter and with their own terminator.

Davies: Presumably one could use a 'helper' virus construct to provide other functions which could then be cut out of the vector. For example gene VI could be removed from the vector leaving space for larger alien inserts, if a helper construct was providing the missing gene product.

Hohn: Yes, but the helper virus would recombine with the vector virus at a high frequency.

Goldbach: You mentioned some sensible applications of CaMV as a viral vector. Compared to the *Agrobacterium* system, by which Mendelian transmission of the transfected gene to the progeny is obtained, using CaMV has the disadvantage that one has to infect either with the virus or *Agrobacterium* again and again. What use will this be?

Hohn: The cloning of the interferon gene is possible in *E. coli*, yeast, insect cells and other systems. If one clones an interesting gene in *E. coli* one needs good purification procedures to extract the protein product and remove bacterial toxins. If it is cloned in an edible plant one might be able to use crude extracts.

Goldbach: That also holds for the *Agrobacterium* system. What advantage does the CaMV system have?

Hohn: A much greater amplification of the gene could be achieved with the virus system than with the *Agrobacterium* system. Transcription is very strong and it may be that we can also improve translation—by deleting the leader sequences, for example.

Harrison: Are you sure there is no integration of DNA into the plant genome when monocots are exposed to 'agroinfection'?

Hohn: We cannot answer this question because the infection is happening at one single cell. There might be an integration event in this cell from which some viruses shed off, but that virus could also recombine out directly from the transferred DNA. To study this question we have to apply agroinfection to tissue culture systems.

Harrison: How could this technique be used to improve virus resistance?

Hohn: Agroinfection has already been used to make transgenic plants containing certain virus gene products, which provide cross-protection.

Davies: It might be interesting to do the type of experiment that was done by de la Peña et al (1987) in Cologne with rye plants. They injected DNA near the flower (in young tillers) at the right time of development and obtained transgenic plants. A combination of the 'agroinfection' maize streak construct and their sub-floral injection technique might enable one to take the transformation to seed.

Hohn: Another approach to obtain transgenic plants would be to put a transposon, such as the mu transposon, into part of the geminivirus genome. That would make the virus transpose and integrate, and then one might get transformed tissue which could be worked on further. The third approach is to improve the tissue culture systems.

Davies: It is interesting that, at least in some geminiviruses like cassava latent virus, the coat protein can be almost entirely deleted and replication and symptoms still occur, but insect transmission does not. Replacement of the coat protein would be an interesting experiment to do with this type of construct.

Nishiguchi: What about the gene product of open reading frame II which is responsible for aphid transmission? Do you have any information about the non-transmissive mutant of CaMV?

Davies: There are several deletion mutants. In those cases it is apparent that if a large proportion of the gene is deleted the product is not obtained. One of the natural isolates, the Campbell isolate, was thought for some years not to produce the gene II protein at all. It seems that it is produced in very low amount—it can be detected by a sensitive Western blotting technique. Therefore transmission or non-transmission is presumably determined by an amino acid difference in those proteins, not the absence or presence of the protein. We have done *in vitro* hybrid experiments exchanging small restriction fragments within the gene II from transmissible to non-transmissible and vice versa. We narrowed the relevant amino acid sequence down to a small region (Woolston et al 1987). When we compared the sequences there was only one amino acid difference in that region. Therefore the difference between transmission and non-transmission can be accounted for by one amino acid change in the gene II protein but we do not know whether that is because the change alters the secondary and tertiary structure of the protein or whether it is in the binding site which interacts with the virus or insect.

White: Have you transformed tobacco plants with your maize streak virus/ *Agrobacterium* construct?

Hohn: No, but we have made transgenic tobacco plants with cauliflower mosaic virus: they have the complete virus genome but don't produce virus.

Sela: Have you transformed tobacco plants to introduce interferon-making virus?

Hohn: No, we have not.

Zaitlin: Do you know what open reading frame VII does?

Hohn: It is definitely not a necessary gene, but it is translated. It codes for a basic protein. Its presence alone results in a regulatory action—it inhibits translation of the following reading frames in *cis*.

Bol: In your transgenic plants containing the cauliflower mosaic virus DNA do you get a transcript that is translated to some extent?

Hohn: There is a transcript and virus is produced in *Brassica*.

References

de la Peña A, Lörz H, Schell J 1987 Transgenic rye plants obtained by injecting DNA into young floral tillers. Nature (Lond) 325:274–276

Woolston CJ, Czaplewski LG, Markham PG, Goad AS, Hull R, Davies JW 1987 Location and sequence of a region of cauliflower mosaic virus gene 2 responsible for aphid transmissibility. Virology, in press

Final general discussion

Durability of resistance

Harrison: I would like to stimulate discussion about the durability of resistance and the ways in which viruses might overcome resistance. Pelham et al (1970) looked at the effects of growing tomato varieties which contained (R) or lacked (S) the *Tm-1* gene for resistance to tomato mosaic. When R varieties were grown for one or more years a new virus strain, strain 1, appeared which caused obvious disease in these varieties and strain 0, the virus strain that was previously prevalent, became much less so. However, strain 1 lacked the biological fitness of strain 0, and when S varieties were grown again strain 0 rapidly reappeared. Thus, strain 1 was prevalent only when the appropriate selection pressure was applied.

With other viruses, it is quite often the case that resistance-breaking strains exist, such as the B strain of potato virus X, but resistance to the virus may nevertheless be quite durable in agriculture. Dr Fraser tells us that there is a strain of TMV able to overcome the resistance conferred by the *N* gene. However, it has not become prevalent, despite the extensive cultivation in some areas of tobacco varieties which contain the *N* gene. Raspberry cultivars with immunity to common strains of raspberry ringspot virus are another example—a resistance-breaking strain has been found but it has not become common. It appears that this strain is not efficiently transmitted through seed, which is one way the virus survives the winter. The resistance-breaking strain therefore has a biological disadvantage which impairs its ability to become prevalent. What experiences have other people had of the durability of virus resistance (controlled either by single genes or by multiple genes) under agricultural conditions? Is there any consensus on the durability of different kinds of virus resistance?

Duffus: One of the most durable resistances has been in sugar beet to the beet curly top virus (McFarlane 1969). The original selections were made 60 years ago. The varieties that were originally selected are still resistant to mild strains of curly top. For the curly top virus, unlike many of the other viruses we have discussed, there seems to be no cross-protection mechanism. I am not sure whether this has anything to do with the lasting resistance. Over the years, the virus strains have increased in severity although the resistant varieties of plants that have been developed always react in the same sequence: severe virus strains are slightly more severe on the early developed resistant varieties than on the later developed resistant varieties. Little is known about the genetics of this system, except that it is multigenic resistance. The mechanisms of that

196

resistance have not been extensively studied, but we do know that two main phenomena are involved. Disease resistance is associated with lower virus concentrations in infected resistance plants and there is a greatly increased incubation period in resistant varieties. The longer incubation period seems to be the main determinant of resistance. If infection is delayed for four to six weeks in the early stages of growth, then the resistance factors seem to take charge—infection with the resulting long incubation period results in relatively little crop loss.

Another similar resistance has been developed for beet western yellows virus in sugar beet (Lewellen & Skoyen 1984). Again there is an apparent lack of cross-protection in sugar beets for the beet western yellows isolates. This resistance has been known for only 20 years but it also seems to be durable, and is multigenic.

Harrison: So resistance is holding up against the original strain. However every time you introduce a new sugar beet cultivar which is slightly more resistant to the curly top virus, you select virus isolates from the population of curly top strains that are more damaging in the old varieties than the strains that were common before. These newly selected strains cause only moderate damage in the new varieties. Therefore you are shifting the population of curly top strains to a more aggressive type.

In this kind of system there seems to be strong selection pressure for more aggressive virus strains. I don't know if these aggressive strains reach a greater concentration in plants, but the lack of cross-protection between the strains allows the more aggressive ones to become established, even in plants already infected with a milder strain.

Duffus: There seems to be some very intricate biology involved. The beet leafhopper is a sun-loving insect and is attracted to plants that are in the open, and are stunted. Therefore there is an increase in exposure of the insects to severe strains because the leafhoppers avoid heavy foliage. Hence, plants infected with mild strains attract less leafhoppers than plants that are severely affected.

Davies: Is it known if the types of resistance you have described, and also the long incubation period, are the same if you mechanically inoculate plants, or do they only apply to leafhopper transmission? Does resistance occur at the insect feeding level rather than by virus replication changes?

Duffus: The mechanical transmission experiments have not been done because mechanical transmission of this virus is virtually nil.

Davies: But it is possible to achieve mechanical transmission by injection or 'pricking' into the crown of young plants.

Duffus: There are plenty of opportunities for leafhoppers to feed on the plant. These results involve more than resistance to infection. The plants become infected, but it takes the virus a lot longer to reproduce and to produce symptoms.

Harrison: Dr Duffus now has a series of strains of curly top of increasing aggressiveness, isolated over several years. The time seems ripe for a molecular analysis of the genomes of these isolates because that might help us to understand how viruses evolve and develop the ability to overcome resistance.

Loebenstein: Another system in which the resistance has been breaking down in the last five to seven years is the transmission of TMV in tomato and sweet pepper seeds. For years it was thought that the virus did not go through the embryo into the next generation. A few virus strains have now been found in different countries which are transmitted through the embryo. With sweet peppers at least, the selection pressure was not strong, because the varieties of sweet peppers that had been grown had the gene for hypersensitivity and the virus did not spread in the plant. Suddenly strains of virus have appeared in the Mediterranean, in Italy and Yugloslavia, which are transmitted through the seed at a very high rate.

Harrison: In tomato this might be because of the practice introduced a few years ago of treating the seed to remove the TMV contaminating the testa which would otherwise have been transmitted to the roots of seedlings during transplanting. Now that this source of TMV has essentially been removed, only the relatively small amount of embryo-infecting TMV remains. Therefore a TMV strain which is more prone to infect the embryo might be selected.

Loebenstein: That could be an explanation for tomatoes, but not for sweet peppers. For the last 10 years, sweet pepper varieties have had a gene for hypersensitivity and the virus did not become systemic in the plant. Therefore, there was no need to remove contaminating virus by any treatments of the seed. Nevertheless these new strains have arisen. Where did they come from? There was not a selection pressure by the geneticist or in the field.

White: Is it not possible that, if this resistance involves a localization reaction, a mutation may occur actually within the localization and the mutant virus may escape and infect the rest of the plant? A mutation occurring within the infection is the one drawback with localization reactions. If resistance is broken down and the virus manages to spread throughout the plant, it is likely to do so very efficiently, and may even enter the embryo.

Harrison: It is not clear why resistance to some viruses breaks down more rapidly than that to others. We have to interest molecular virologists in investigating what is happening, so that the molecular changes underlying the biological ones are determined.

Fraser: We have to expect that different resistance genes will impart different durabilities. *Tm-1* alone is useless for protecting tomato against TMV. The other end of the spectrum in tomato is *Tm-2²* which has been very successful against TMV for about 10 years. However, several research groups now have virus isolates which overcome *Tm-2²*, and if these were to become widespread in the glasshouse tomato industry, the whole basis of our protection against tobacco mosaic virus would be in shreds.

We know of many resistance-breaking isolates. My statistical survey (Fraser, this volume) suggests that 60% of resistance genes have definitely been overcome. However, one has to consider the geographical isolation of some of the virulent isolates. Resistance genes that have been overcome can be useful in other parts of the world, which means that resistance is still a very valuable strategy. The lesson of the *N* gene is that we have to expect any gene, no matter how long it has been durable, eventually to be overcome.

Dr Harrison has told us that the strain 1 isolates of TMV are replaced by strain 0 isolates in tomato if one starts growing susceptible varieties. We have characterized a series of spontaneous mutants and artificial mutants to virulence against *Tm-1* and virulence is associated with a loss in strength of the virus: it either multiplies less well or causes milder symptoms on the susceptible host. In this case virulence does seem to have a price. However, we characterized natural isolates of strain 1 type from glasshouse cultivation, they multiplied just as well and caused just as severe symptoms as the strain 0 isolate. Therefore, although it seems that mutation to virulence may have a price, susequent mutation may increase the fitness of the isolate.

Harrison: That is a good point. Dr van Vloten-Doting commented earlier that if you change one nucleotide in a virus genome, this may lead to a loss of aggressiveness or a loss of ability to survive in nature. Presumably this is how virus strains arise: they are mostly distinguished by a series of nucleotide changes rather than a single mutation.

van Vloten-Doting: We compared a particular isolate of AlMV, which could infect beans systemically but was derived from a parent strain that induced local lesions on beans, to the parent strain and to a natural isolate that was also systemic in beans. When we compared the mutant to the parent we saw over 60 mutations; to our surprise and delight many of those mutations are also present in the natural isolate. However, some mutations are lacking and there are additional mutations. This mutant is certainly not a contamination with the natural isolate. Apparently isolates having the same phenotype share the same genotype to some extent. The problem is to find out which mutation is determining what characteristic.

Harrison: Dr Fraser suggested that it was a good strategy to breed tomatoes with three different major genes for resistance to TMV to obtain more durable resistance. Nobody would want to disagree with that. However, one might contrast that situation with the larger number of genes involved in resistance to beet curly top virus, each of which may have a smaller effect. This type of genetic control does not seem to be a recipe for durable resistance. In fact, it is necessary to continue to breed against increasingly aggressive forms of the virus.

Duffus: There is another situation, turnip mosaic virus resistance or immunity in lettuce (*Lactuca sativa*) (Zink & Duffus 1969). This system has been tested with strains of turnip mosaic virus from all over the world and no one has yet found an isolate able to attack those resistant varieties. It seems that a single gene is responsible and that the immunity is absolute.

Fraser: If the polygenic strategy is considered in the light of the gene-to-gene interaction concept which I outlined in my paper (Fraser, this volume), the question is raised of how many genes for virulence can be loaded into the viral genome. Presumably some other important virus function is altered at each change. Therefore there is probably a limit to the number of virulence functions that can be successfully incorporated. That is the attraction of the polygenic strategy. It is clear from virus genetics that it is possible to get more than one or two virulence functions into the virus, but there may be a limit.

Hohn: Can we use the model of the influenza virus of the animal system? I suggest that the resistance factor might be similar to a type of 'antibody', by being an inherited factor that binds to some part of the virus. The influenza virus overcomes the resistance by mutating the site that interacts with an antibody. The plant virus overcomes resistance by changing the gene interacting with the plant resistance factor. The hosts produce new antibodies, or new plant lines, as a counter measure. It could be the same interaction, the same virulence function, which is modified repeatedly and there might not be a need for more than one of these functions for a given virus. The same function may be changed, perhaps even in repeated cycles as was suggested for influenza.

Fraser: There is still the possibility of a constraint. The virus may be driven to mutate in different directions which are mutually incompatible. My example is that we do not have any strain of TMV which can overcome both *Tm-2* and *Tm-2²* yet. That does not prove that such a strain will not eventually develop but it may be a very difficult or impossible evolutionary step for the virus.

Sänger: The influenza virus is not a good model, because there are eight segments of the genome and there is an exchange between different isolates from different animal species.

Hohn: That could happen in plants too, for example by pseudorecombination of multipartite viruses.

Harrison: There are two separate kinds of variation in influenza virus. One is the antigenic shift that occurs when pseudorecombination leads to a reassortment of the genome segments of two different isolates. The other is the antigenic drift found as a particular region of the genome part accumulates mutations in the way that Dr Hohn has explained. In such a system, resistance would not be absolute and the question becomes how well a virus replicates, rather than whether or not it replicates.

Dodds: It should be remembered that plant varieties selected for resistance in one geographical region may be of little value in another region, where the virus strains may be sufficiently different to overcome the selected resistance. That has happened with pepper virus resistance in the USA. Varieties selected for Texas conditions were promising, but when the most desirable Texas lines were planted in the field in California they became infected by the same viruses. No obvious selection pressures were being applied—effective strains were there waiting for these supposedly resistant plants.

Harrison: We are all delighted when we achieve resistance in a laboratory, but what really matters is how well the plants perform when they are grown as crops. That is a question which will require more attention in the future.

References

Fraser RSS 1987 Genetics of plant resistance to viruses. In: Plant resistance to viruses. Wiley, Chichester (Ciba Found Symp 133) p 6–22

Lewellen RT, Skoyen IO 1984 Beet western yellows can cause heavy losses in sugar-beet. Calif Agric 38:4–5

McFarlane JS 1969 Breeding for resistance to curly top. J Inst Sugar Beet Res 4:73–83

Pelham J, Fletcher JT, Hawkins JH 1970 The establishment of a new strain of tobacco mosaic virus resulting from the use of resistant varieties of tomato. Ann Appl Biol 65:293–297

Zink FW, Duffus JE 1969 Relationship of turnip mosaic virus susceptibility and downy mildew (*Bremia lactucae*) resistance in lettuce. J Am Soc Hort Sci 94:403–407

Summary

Pathogenesis

Fraser: It seems appropriate to begin my summary of the progress in understanding pathogenesis with a historical perspective by referring to the *N* gene. It was timely for Dr Zaitlin to point out that we do not know if the susceptible allele is a null allele or if it is active (Dunigan et al, this volume). Some modern cytogenetic investigations on tobacco might be very informative. We all sympathize with Dr Zaitlin's herculean attempt to isolate the product of the *N* gene. Perhaps transposon mutagenesis will be a more promising approach to isolation of the gene itself. Our understanding of the fundamental aspects of resistance mechanisms will be furthered by obtaining genes, sequencing them and trying to solve their function.

In my paper (Fraser, this volume) I mentioned gene-for-gene relationships because I find this a useful concept when thinking about virus–plant interactions. In discussion it was apparent that some people are not too happy with the concept. I would like to stress that it is a simplifying idea rather than the whole picture. However, there is no doubt that we do have gene-for-gene relationships between resistance in the plant and virulence in the virus.

The virulence gene presumably fulfils some other function, as far as viral pathogenesis is concerned, and, to some extent, we need to regard the virus population as a multicopy gene family when we study its evolution. We discussed possible ways to characterize virulence in the virus. We discovered that the sequencing approach would lead to an embarassment of information that would be very difficult to interpret. I think we should be looking at the functions specified by the virus rather than the nucleic acid sequence, and should try to relate alterations in particular functions to alterations in interaction with a resistance gene product.

We have considered possible modes of resistance gene action, largely in the context of the *N* gene. Two experimental approaches were described. One is the descriptive study of the array of changes which occur when the resistance mechanism is induced, such as that undertaken by Fritig et al (this volume). The other is to look directly for some antiviral activity by an assay which involves the virus itself.

We have discussed two examples of the latter: Dr Sela's study of the antiviral factor (AVF) (Sela et al, this volume) and Dr Loebenstein's study of the inhibitor of virus replication (IVR) (p 116). One of the problems associated with these experiments is that we are not sure which stage of the virus replicative cycle is inhibited. We require an understanding of the mode of action of both these putative inhibitors. We may need to move towards more specific assay systems; ultimately using *in vitro* assays and looking at the inhibition of replication or translation. The parallels between AVF and interferon are intriguing, but the evidence for interferon action against plant viruses is at best equivocal. I would like to reiterate Dr Sänger's suggestion (p 114) that AVF should be purified and sequenced, and its specific action established.

Most of the changes that occur during virus localization do so after necrosis. A clear picture has emerged of a cascade reaction in which the virus eventually induces an array of changes, such as alterations in cell walls and the production of many enzymes. Many, if not all, of these metabolic changes may be involved in resistance, which is widespread and can be effective against fungi and bacteria. We do not seem to be sure whether, or how, these resistance mechanisms operate against viruses.

The particular item of information that we are lacking is the top of the cascade where the initial recognition event occurs and where the various processes are switched on. We have useful data on the subsequent processes and how these changes can induce various aspects of resistance, such as cell wall changes and antifungal substances.

Progress has been made in the study of the pathogenesis-related (PR) proteins. Dr Bol told us about his elegant work on the characterization of the genes (Bol et al, this volume). This, together with Dr Fritig's work, has produced many intriguing results. In particular, eight of the sixteen PR proteins in tobacco have now been assigned a function, which is reassuring. The generality of the resistance that is produced is illustrated by the PR proteins that confer resistance against fungi and bacteria by chitinase and β-1,3-glucanase activity. These results, however, fail to answer the question of the possible role of PR proteins in resistance against viruses. The interesting tests will be with transgenic plants, to see if we can actually find resistance. There is also an interesting paradox in that we have assigned functions for half of the PR proteins, and some of the others will be resolved quickly, but the functions of the proteins PR-1a, -1b and -1c, which were the first to be discovered, are still unknown.

Several questions remain open. I think we have established that localization of virus occurs before and without necrosis. Why do we *get* necrosis in so many examples of localization? How is it that we have one or two examples where a virus can be localized without necrosis? What about tobacco necrosis virus? It almost always causes a necrotic infection. Does every plant have a gene for resistance or is it something intrinsic to the virus which induces necrosis?

We have also discussed some other types of resistance. The information that we have been given on the 30K transport protein is intriguing (Nishiguchi & Motoyoshi, this volume) and it will be fascinating to see how that may be related to the mechanisms of resistance which operate by preventing virus spread, especially in tomato.

Finally, Dr Bruening's work on the Arlington cowpea protease inhibitor (Bruening et al, this volume) provides a very nice biochemical explanation of resistance and of virulence in the virus. It would be interesting to have more detail on the specificity of the protease inhibitor and its interaction with the protease. Dr Bruening told us how he discovered the resistance in Arlington— it is not present in hundreds of other cowpea lines—when he ran out of seed and borrowed some from a colleague. Everyone should have such a stroke of good fortune at least once!

Genetic engineering approaches to plant resistance

van Vloten-Doting: We can distinguish two approaches to the genetic engineering of virus resistance in plants. The first possibility is to generate resistance to one particular virus. However, plant breeders are often of the opinion that the value added to seed by the introduction of resistance to one particular virus is insufficient for this approach to be economically viable. The other possibility is to try to fulfil the breeder's dream by producing plants resistant to all pathogens. This, of course, is very difficult!

First let us look at the possibility of introducing resistance to one particular virus. At present we can use either plant-coded resistance genes or parts of the viral genome to interfere with virus replication.

A fair number of resistance genes have been identified in crop species and there are probably many more resistance genes present in wild species. Once a gene is identified, how can it be isolated? The most promising approach is by transposon tagging. In crop species, where (as yet) no transposons are present, restriction fragment length polymorphism (RFLP) analysis may be used. The non-host type of resistance gene, discussed by Dr Fraser (this volume), is interesting. Unfortunately, genes of this type will probably often be recessive, making gene transfer useless when the dominant gene is still present. We therefore have to produce a system for gene substitution. This requires homologous recombination of the new gene with the resident gene. Some non-host resistances may be of the new preferred type that acts against several viruses. However, in the light of the findings of Bruening et al (this volume) that Arlington cowpea has resistance to two unrelated viruses based on two different mechanisms, one should be aware that the different resistances may not be related.

Instead of using information from the host plants, it is possible to use genetic information from a virus. Dr Beachy has elegantly demonstrated that if one puts the information for virus coat protein into tobacco plants, one can measure a quantitative effect on protection against viruses (Beachy et al, this volume).

Surprisingly this protection was seen not only against the virus from which the genetic information was derived but also against other unrelated viruses. The results are not yet complete: one more control experiment should be done, to show that it really is the coat protein and not the viral RNA that is important. This could be achieved by making a frame-shift mutation keeping the same amount of viral RNA in the absence of the coat protein. Plants carrying this construct should not show virus resistance.

This work with the coat protein brings us closer to the goal of general resistance because the coat protein may help to curtail several viruses. Dr Beachy showed that when TMV coat protein is present, other viruses replicate to a lower level in the initially inoculated leaf and are prevented from spreading to the upper leaves. It is possible that when the amount of virus induced in the primarily infected leaves is lowered, the natural resistance of the plant is sufficient to prevent spreading.

When we considered the role of antisense RNA, Dr van Emmelo (p 181) showed that care must be taken and that it may be necessary to modulate and modify the antisense molecules before use. Dr Baulcombe pointed out that the satellite RNA approach is applicable to few viruses at present (Baulcombe et al, this volume). It is fascinating that we can use satellite RNA but we do not know how it works: tomato aspermy virus replicates as if no satellite RNA was present but the infected plants do not show any symptoms. It may be appropriate to combine two different protection methods so that if one fails, the other still works.

What strategies can we use to obtain a more general resistance in plants to pathogens? First one can activate the natural defence system. This must be carefully regulated, because the *Nicotiana* hybrids that constitutively produce defence proteins are not very healthy. Often a defence mechanism will result in cell death.

Another approach would be to apply external elicitors. The use of cell wall fragments or antiviral factor (AVF) may be promising. One might even try to use beneficial bacteria living on the plants to produce elicitors continuously. If the elicitor induces part of the defence reaction without necrosis, nature may do our work for us.

What do we know about other genes that protect plants against viruses, and is it useful to try to transfer them to crop plants? Dr Fritig told us that some of the PR proteins have enzymic activity (Fritig et al, this volume). It may be useful to put them in plants and see what happens. However, the cascade of metabolic changes induced is so complex that it might be too ambitious to expect a transgenic plant with just one or two of these genes to be highly resistant.

What are the most important problems and where should we work hardest? The issue of durability of resistance is essential. How long will the engineered virus resistance last in nature? We have to do the experiments to find out, but the effects and the side-effects of this type of experiment should always be considered. To what extent will other viruses and other pathogens spread? The answer to this question will differ with the type of the experiment.

We need to do a lot of transposon tagging and RFLP analysis to obtain resistance genes, and methods for gene substitution will have to be worked out. It is also necessary to address the question raised by Dr Fritig—what is the initial event that activates the defence mechanisms? The transport mechanism is by far the most interesting area that requires investigation at the moment. We should try to understand how viruses move and how we can contain them.

*   *   *

Harrison: In conclusion, I should like to say that this has been a very good symposium and a very timely one. Many people have contributed to its success and we are all indebted to Dr Karl Maramorosch for providing the germ of the idea for such a useful meeting.

References

Baulcombe DC, Hamilton WDO, Mayo MA, Harrison BD 1987 Resistance to viral disease through expression of viral genetic material from the plant genome. In: Plant resistance to viruses. Wiley, Chichester (Ciba Found Symp 133) p 170–184

Beachy RN, Powell Abel P, Nelson RS, Register J, Tumer N, Fraley RT 1987 In: Plant resistance to viruses. Wiley, Chichester (Ciba Found Symp 133) p 151–169

Bol JF, Hooft van Huijsduijnen RAM, Cornelissen BJC, van Kan JAL 1987 Character-ization of pathogenesis-related proteins and genes. In: Plant resistance to viruses. Wiley, Chichester (Ciba Found Symp 133) p 72–91

Bruening G, Ponz F, Glascock C, Russell ML, Rowhani A, Chay C 1987 Resistance of cowpeas to cowpea mosaic virus and to tobacco ringspot virus. In: Plant resistance to viruses. Wiley, Chichester (Ciba Found Symp 133) p 23–37

Dunigan DD, Golemboski DB, Zaitlin M 1987 Analysis of the *N* gene of *Nicotiana*. In: Plant resistance to viruses. Wiley, Chichester (Ciba Found Symp 133) p 120–135

Fraser RSS 1987 Genetics of plant resistance to viruses. In: Plant resistance to viruses. Wiley, Chichester (Ciba Found Symp 133) p 6–22

Fritig B, Kauffmann S, Dumas B, Geoffroy P, Kopp M, Legrand M 1987 Mechanisms of the hypersensitivity reaction of plants. In: Plant resistance to viruses. Wiley, Chichester (Ciba Found Symp 133) p 92–108

Nishiguchi M, Motoyoshi F 1987 Resistance mechanisms of tobacco mosaic virus strains in tomato and tobacco. In: Plant resistance to viruses. Wiley, Chichester (Ciba Found Symp 133) p 38–56

Sela I, Grafi G, Sher N, Edelbaum O, Yagev H, Gerassi E 1987 Resistance systems related to the *N* gene and their comparison with interferon. In: Plant resistance to viruses. Wiley, Chichester (Ciba Found Symp 133) p 109–119

Index of contributors

Non-participating co-authors are indicated by asterisks. Entries in bold type indicate papers; other entries refer to discussion contributions

Indexes compiled by John Rivers

Subject index